Neuropsychiatric Systemic Lupus Erythematosus

Shunsei Hirohata
Editor

Neuropsychiatric Systemic Lupus Erythematosus

Pathogenesis, Clinical Aspects and Treatment

Springer

Editor
Shunsei Hirohata
Department of Rheumatology
Nobubara Hospital
Tatsuno, Hyogo, Japan

Department of Rheumatology & Infectious Diseases
Kitasato University School of Medicine
Sagamihara, Kanagawa, Japan

ISBN 978-3-319-76495-5   ISBN 978-3-319-76496-2   (eBook)
https://doi.org/10.1007/978-3-319-76496-2

Library of Congress Control Number: 2018937679

© Springer International Publishing AG, part of Springer Nature 2018
This work is subject to copyright. All rights are reserved by the Publisher, whether the whole or part of the material is concerned, specifically the rights of translation, reprinting, reuse of illustrations, recitation, broadcasting, reproduction on microfilms or in any other physical way, and transmission or information storage and retrieval, electronic adaptation, computer software, or by similar or dissimilar methodology now known or hereafter developed.
The use of general descriptive names, registered names, trademarks, service marks, etc. in this publication does not imply, even in the absence of a specific statement, that such names are exempt from the relevant protective laws and regulations and therefore free for general use.
The publisher, the authors and the editors are safe to assume that the advice and information in this book are believed to be true and accurate at the date of publication. Neither the publisher nor the authors or the editors give a warranty, express or implied, with respect to the material contained herein or for any errors or omissions that may have been made. The publisher remains neutral with regard to jurisdictional claims in published maps and institutional affiliations.

Printed on acid-free paper

This Springer imprint is published by the registered company Springer International Publishing AG part of Springer Nature.
The registered company address is: Gewerbestrasse 11, 6330 Cham, Switzerland

Preface

Thirty-five years have passed since I started on working in the field of rheumatology at the University of Tokyo Hospital in 1982. Thus far, during these 35 years, enormous progress has been achieved in terms of genetics, immunology, diagnosis and treatment of neuropsychiatric systemic lupus erythematosus (NPSLE). In particular, my own discovery of the elevation of cerebrospinal fluid (CSF) interleukin-6 in NPSLE in 1990 has enabled us to make a correct diagnosis of NPSLE until now. On the other hand, as for autoantibodies, both anti-ribosomal P protein antibodies and anti-NMDAR NR2 antibodies have brought us great impact on our understanding of the pathogenesis of diffuse NPSLE.

I was lucky to meet so many patients in the University of Tokyo Hospital, Teikyo University Hospital and Kitasato University Hospital. Despite the remarkable progress mentioned above, there are still unmet needs in the field of NPSLE. Especially, there are still such patients that could not be successfully saved. I always remember one junior high school female student, who died of fulminant brainstem encephalitis that I had never seen before. Every effort to delineate the mechanism of the attack was in vain. However, there must be a definite abnormality leading to the development of such a fulminant disease in NPSLE. Hopefully, the mechanism will be disclosed by younger researchers in the near future. Several issues have been addressed in this book along with the progress in the past three decades by leading specialists in the respective field of NPSLE. It would be my enormous pleasure if this book will make even a tiny contribution to such a great progress.

My special thanks go to all the authors of this book who make great contributions. I would also like to thank everybody who helped me to reach the point of publishing this book, especially Prof. Masahisa Kyogoku who gave me helpful advice on pathology as well as all the patients who trust us. Finally, we also thank Springer US for this exciting opportunity.

Hyogo, Japan   Shunsei Hirohata

Contents

1 **Epidemiology of Neuropsychiatric Systemic Lupus Erythematosus**.................................... 1
Shunsei Hirohata

2 **Genetics** 15
Naoyuki Tsuchiya

3 **Immunopathology of Neuropsychiatric Systemic Lupus Erythematosus**.................................... 29
Shunsei Hirohata

4 **Pathology of Neuropsychiatric Systemic Lupus Erythematosus**.................................... 43
Shunsei Hirohata

5 **Clinical Features** 59
Yoshiyuki Arinuma and Shunsei Hirohata

6 **Cytokines and Chemokines** 77
Taku Yoshio and Hiroshi Okamoto

7 **Diagnosis and Differential Diagnosis** 93
Taku Yoshio and Hiroshi Okamoto

8 **Imaging of Neuropsychiatric Systemic Lupus Erythematosus**...... 113
Yoshiyuki Arinuma and Shunsei Hirohata

9 **Psychiatric Symptoms** 129
Katsuji Nishimura

10	**Treatment of Neuropsychiatric Systemic Lupus Erythematosus**.................................	141
	Tetsuji Sawada	
11	**Promising Treatment Alternatives**...........................	155
	Taku Yoshio and Hiroshi Okamoto	
12	**Prognosis of Neuropsychiatric Systemic Lupus Erythematosus**....	169
	Shinsuke Yasuda	

Index.. 185

Chapter 1
Epidemiology of Neuropsychiatric Systemic Lupus Erythematosus

Shunsei Hirohata

Abstract A variety of neuropsychiatric manifestations are observed in patients with SLE. The American College of Rheumatology (ACR) developed standardized nomenclature and case definitions for neuropsychiatric involvement in SLE (NPSLE) in 1999. One of the problems in the 1999 ACR classification is the inclusion of milder, less specific and more subjective manifestations such as headache, mild cognitive dysfunction and mood disorders, which resulted in an enormous variation in the prevalence between studies. Another critical point of the ACR classification is the lack of a number of other neurological manifestations, such as neuromyelitis optica spectrum disorders and reversible focal neurological deficits mimicking cerebrovascular disease. Steroid psychosis is sometimes a difficult differential diagnosis, but not necessarily an exclusion, of lupus psychosis. CSF IL-6 might be one of the surrogate markers to detect patients with headache, cognitive dysfunction and mood disorders, requiring immunosuppressive therapy.

Keywords Epidemiology · Prevalence · American College of Rheumatology Classification · Mortality

1.1 Introduction

Neuropsychiatric involvement in systemic lupus erythematosus (SLE) is one of the recalcitrant complications of the disease, leading to substantial impairment of quality of life as well as disability [1, 2]. A variety of neuropsychiatric manifestations are seen in patients with SLE. Thus, such complexity has made it difficult to make a correct diagnosis and introduce an appropriate treatment. The American College of Rheumatology (ACR) developed standardized nomenclature and case definitions

---

S. Hirohata (✉)
Department of Rheumatology, Nobuhara Hospital, Tatsuno, Hyogo, Japan

Department of Rheumatology & Infectious Diseases, Kitasato University School of Medicine, Sagamihara, Kanagawa, Japan

© Springer International Publishing AG, part of Springer Nature 2018
S. Hirohata (ed.), *Neuropsychiatric Systemic Lupus Erythematosus*,
https://doi.org/10.1007/978-3-319-76496-2_1

for neuropsychiatric involvement in SLE (NPSLE) in 1999, which has enabled the epidemiological studies to be performed on an equal basis [3].

In this chapter, the overall epidemiological features in NPSLE will be described. Furthermore, the limitations in the ACR nomenclature will be discussed.

1.2 Classification

The first attempt for classification of NPSLE was the inclusion of seizures, psychosis and focal neuropsychiatric events in the preliminary SLE classification in 1971 [4]. In the 1982 ARA revised criteria for SLE only seizures and psychosis were included [5]. In 1985, Harris and Hughes summarized the classification of manifestations of NPSLE in the literature, providing the prevalence of various neuropsychiatric manifestations in several studies (Table 1.1) [2]. In this classification, psychiatric manifestations were classified into 2 categories, including organic brain syndrome and non-organic brain syndrome [2]. Organic brain syndrome, originally created for discrimination of psychiatric disturbances due to physical causes from functional psychiatric disorders, was characterized by impairment of orientation, perception, memory, or intellectual function [6]. Non-organic brain syndrome was characterized by neurosis, depression, psychosis or schizophrenia [7]. Most series of studies reported that psychiatric abnormalities and seizures were the most frequent neuropsychiatric disorders of SLE. In fact, both psychiatric abnormalities and seizures are included in the 1982 revised criteria for the classification of SLE [5].

Table 1.1 The frequency of neuropsychiatric manifestations in selected series of patients with SLE

Clinical manifestations	Authors (study design) [Reference] Gibson and Mayers (retrospective) [1]	Grigor et al. (prospective) [6]
Number of patients studied	80	50
All neuropsychiatric features	51%	50%
Organic brain syndrome	19%	18%
Psychiatric illness	8%	22%
Seizures	20%	14%
Cranial nerve palsies	4%	16%
Stroke	10%	16%
Movement disorders[a]	4%	4%
Myelopathy	3%	–
Peripheral neuropathy	2%	6%
Visual defects	2%	–
Aseptic meningitis	1%	–

[a] Movement disorders include with cerebellar ataxia and chorea

Table 1.2 The American College of Rheumatology nomenclature and case definitions for neuropsychiatric lupus syndromes (1999)

Central nervous system	
Neurologic syndromes	
	Aseptic meningitis
	Cerebrovascular disease
	Demyelinating syndrome
	Headache (including migraine and benign intracranial hypertension)
	Movement disorder (chorea)
	Myelopathy
	Seizure disorders
Diffuse psychiatric/neuropsychological syndromes	
	Acute confusional state
	Anxiety disorder
	Cognitive dysfunction
	Mood disorder
	Psychosis
Peripheral nervous system	
	Acute inflammatory demyelinating polyradiculoneuropathy (Guillain-Barré syndrome)
	Autonomic disorder
	Mononeuropathy, single/multiplex
	Myasthenia gravis
	Neuropathy, cranial
	Plexopathy
	Polyneuropathy

All types of seizures, including generalized seizures and focal seizures, may occur [2]. Organic brain syndrome, non-organic brain syndrome and seizures were the most frequent disorders [2]. It should be noted that seizures may arise alone or sometimes be associated with psychiatric abnormalities [8, 9].

The 1999 ACR nomenclature and case definitions for neuropsychiatric involvement in SLE consist of 12 central nervous system (CNS) manifestations and 7 peripheral nervous system (PNS) manifestations, which are considered to be related with SLE (Table 1.2) [3]. The 1999 ACR nomenclature and case definitions provide diagnostic criteria, exclusion criteria to rule out neuropsychiatric events unrelated to SLE, associations to consider concomitant or pre-existing comorbidities, a set of recommendations to confirm each neuropsychiatric events as appendix [3].

Since the term "organic brain syndrome" is sometimes misleading, it has been replaced by the term "diffuse psychiatric/neuropsychological syndromes" in the 1999 ACR nomenclature and case definitions [3]. Thus, organic brain syndrome and non-organic brain syndrome were reformed into 5 domains of diffuse psychiatric/neuropsychological manifestations, including acute confusional state, anxiety disorder, cognitive dysfunction, mood disorder and psychosis [3]. Among these, acute confusional state is the most severe manifestation, requiring extensive immunosuppressive therapy and sometimes resulting in poor prognosis [3, 10]. Acute organic brain syndrome and chronic organic brain syndrome in the previous classification are considered to correspond to acute confusional state and cognitive dysfunction, respectively, in the 1999 ACR criteria [2, 3]. Diffuse psychiatric/ neuropsychological manifestation is sometimes called as lupus psychosis.

The frequency of each manifestation is variable depending mainly on the nature of the studies. Thus, headache is more frequent in the prospective studies, since milder forms of headache might be overlooked in the retrospective studies. Of note, in the study by Steup-Beekman et al., which included the patients referred for the purpose of evaluation on MRI scans, the frequencies of cerebrovascular disease, headache and cognitive dysfunction are much higher than those in other studies [11].

It should be pointed out that the ACR nomenclature and definitions contain such manifestations that might consist of different degrees of severity. For example, cognitive dysfunction comprises of manifestations from mild subclinical deficits to severe dementia, the former being more commonly observed even in the general population [12].

1.3 Demographic Features of NPSLE

1.3.1 Prevalence

A number of studies have explored the demographic features of NPSLE. The reported overall prevalence of NPSLE in the previous studies ranges widely between 14 to 95%, even after the introduction of 1999 ACR nomenclature and definitions [13–16]. A number of factors are involved in such variation between studies, including the design of the study, differences in selection criteria and ethnic differences in the studied population. In general, studies with a larger number of patients would result in less selection bias. However, even in studies with the largest cohorts of SLE patients, the prevalence of NPSLE was variable with a range from 19% to 57%, although the variation was smaller than studies with smaller number of patients [12, 17, 18]. It should be remembered that less specific manifestations such as headache, cognitive dysfunction, mood disorders tend to be more commonly observed even in the general population [12]. The inclusion or exclusion of minor neuropsychiatric manifestations, such as mild cognitive dysfunction detected only by a structured battery test, would result in a significant variability of the prevalence [14, 17].

The results of a meta-analysis of 5057 SLE patients have revealed the prevalence of NPSLE of 44.5% in prospective studies and 17.6% in retrospective studies (Table 1.3) [19]. In the subanalysis of the 10 prospective and elicited studies of higher quality (2049 patients) using the random effects model, the prevalence of NP

Table 1.3 Prevalence of neuropsychiatric manifestations according to study design

Clinical manifestations	All patients (n = 5057) Number Prevalence	Prospective/ elicited studies (n = 2049) Prevalence	Retrospective studies (n = 3008) Prevalence	p value
Headache	617 12.2%	23.3%	4.7%	p < 0.001
Mood disorder	376 7.4%	14.9%	2.3%	p < 0.001
Seizure disorder	356 7.0%	8.0%	6.4%	p = 0.03
Cognitive dysfunction	334 6.6%	13.9%	1.6%	p < 0.001
CVD	255 5.0%	7.2%	3.6%	p < 0.001
Psychosis	165 3.3%	3.9%	2.8%	p = 0.03
Acute confusional state	155 3.1%	3.9%	2.5%	p = 0.004
Anxiety disorder	121 2.4%	5.3%	0.4%	p < 0.001
Polyneuropathy	76 1.5%	3.0%	0.5%	p < 0.001
Cranial neuropathy	49 1.0%	1.7%	0.5%	p < 0.001
Aseptic meningitis	46 0.9%	0.3%	1.3%	p < 0.001
Myelopathy	45 0.9%	1.0%	0.8%	p = 0.40
Mononeuropathy (single, multiplex)	45 0.9%	1.5%	0.5%	p < 0.001
Movement disorder	36 0.7%	0.9%	0.6%	p = 0.25
Demyelinating syndrome	17 0.3%	0.3%	0.3%	p = 0.96
Myastenia gravis	8 0.2%	0.4%	0%	p < 0.001
AIDP(GBS)	4 0.1%	0.1%	0.1%	p = 0.70
Autonomic disorder	4 0.1%	0.1%	0%	p = 0.16
Total	2709 28.5%	44.5%	17.6%	p < 0.001

[a] Plexopathy was omitted due to the presence of no patient [19]

SLE patients was estimated to be 56.3%, and the most frequent neuropsychiatric were headache 28.3%, mood disorders 20.7%, cognitive dysfunction 19.7%, seizures 9.9%, and cerebrovascular disease 8.0%, although there was a significant variation between studies [19]. In a recent study with lupus Canadian cohort including 1253 patients with mean disease duration of 12 years, the prevalence of NPSLE varied depending on the definition used: 6.4% in a group of NPSLE with seizures or psychosis by ACR classification criteria; 38.6% in a group with seizures, psychosis, organic brain syndrome, cerebrovascular accident, cognitive dysfunction, headache, cranial or peripheral neuropathy, and transverse myelitis with minor nonspecific manifestations such as mild depression, mild cognitive impairment, and electromyogram-negative neuropathies [20]. Thus, one should assume that it is difficult to obtain the exact figure of the prevalence for "NPSLE".

1.3.2 Age of Onset and Ethnicity

Although SLE is more frequent in females of child-bearing age, a similar gender-related tendency is not so evident in NPSLE [21]. As for ethnicity, Hispanics, African descendants and Asians were found to develop NPSLE more frequently than Caucasians [22–25]. Notably, Asian patients also have more severe disease activities of SLE, and possibly NPSLE, compared with Caucasians [22].

NPSLE may frequently occur early in the course of SLE. Thus, neuropsychiatric manifestations develop within the first or second year after diagnosis of SLE in about 50% of the patients [15]. Notably, some reports disclosed that 28–40% of NPSLE-related events occur before or around the time of the diagnosis of SLE [26, 27]. Thus, one needs to keep it in mind that NPSLE might be the initial manifestation of SLE.

1.3.3 Risk Factors

There have been at least 3 major risk factors for NPSLE acknowledged by 2010 EULAR recommendations [10] as well as by other studies. First, generalized SLE disease activity and cumulative damages have been shown to be associated with an increased risk of seizures and severe cognitive dysfunction [28–31]. Second, previous or concurrent occurrences of major NPSLE events, particularly stroke and seizures, have been found to predict similar recurrent events in the future [32, 33]. Third, anti-phospholipid antibodies were associated with cerebrovascular disesse (CVD) [29, 32], seizures [17, 28, 34], myelopathy [35], movement disorders [17] and moderate to severe cognitive dysfunction [17, 30, 36]. The results in the prospective study using the SLICC inception cohort disclosed that lupus anticoagulant (LAC) at baseline was associated with the development of cerebral thrombosis and that anti-ribosomal P antibodies were found to be a risk factor for the occurrence of psychosis [37]. Therefore, it is suggested that the expression of certain autoantibodies due to the SLE disease activity might most likely predict the risk for NPSLE.

1.4 The Limitations of 1999 ACR Nomenclatures and Definitions

First of all, as mentioned above, it should be pointed out that in the 1999 ACR nomenclature and definitions milder forms of neuropsychiatric manifestations are included, such as headache, minor cognitive dysfunctions and subtle mood disorders [3]. In cases with these manifestations, there are a large number of patients who do not need immunosuppressive therapy and usually improve only with a conservative, symptomatic or supportive therapy. However, there are a fraction of patients with cognitive dysfunction or mood disorder who require immunosuppressive therapy [10, 38, 39]. Needless to say, the ACR criteria do not refer to the discrimination of such a fraction of patients. In addition, for lupus headache, included in SLEDAI-2 K [40] and in British Isles Lupus Assessment Group 2004 index [41], corticosteroids (1 mg/kg/day prednisone) proved to be more effective than conventional antimigraine therapy [42]. Thus, it is mandatory to discriminate patients with these manifestations who need immunosuppressive therapy from those who require only supportive therapy. In fact, there are some studies that claimed to exclude milder, less specific and more subjective NP syndromes such as headache, mild cognitive dysfunction and mood disorders along with peripheral neuropathy in order to increase the specificity for SLE [13].

On the next topic, CVD is defined as neurologic deficits usually due to arterial insufficiency or occlusion, venous occlusion disease, or hemorrhage [3]. Most of the patients with CVD have evidence of thrombosis due to antiphospholipid antibodies and suffer from irreversible neurological damages, which do not respond to immunosuppressive therapy [43]. Meanwhile, recent studied have disclosed the presence of patients showing neurological focal deficits along with MRI abnormalities which respond to immunosuppressive therapy to improve almost completely [44]. Such reversible focal neurological deficits have not been included in the 1999 ACR nomenclature [3]. Thus, one cannot help classifying such disorders as CVD.

Finally, in the 1999 ACR nomenclature, myelopathy does not discriminate neuromyelitis optica spectrum disorders (NMOSD) caused by anti-AQP-4 antibodies from transverse myelitis due to vasculitis [45]. These 2 conditions result in different prognosis, and therefore need to be differentiated. Even the ACR criteria encourage the classification of Devic's syndrome (currently NMOSD) as both myelopathy and demyelinating syndrome. It is quite doubtful that NMOSD most suitably classified as demyelinating syndrome in the 1999 ACR nomenclature and definitions [3]. Definitely, reclassification of demyelinating syndrome and myelopathy would be mandatory in the future.

1.5 NPSLE in Childhood

In general, patients with pediatric SLE has been associated with greater severity and poorer prognosis than those with adult SLE [46, 47]. There have been few studies comparing clinical features and prevalence of NPSLE in children and adults. Overall, neurological manifestations are more severe in children, leading to permanent damages at higher rates than adults [48–50]. Previous studies in children over a 6-year study period revealed that neurological manifestations were more common than lupus nephritis (95% versus 55%) [46]. The common NPSLE manifestations in this longitudinal study included headaches in 72% of children, mood disorder in 57%, cognitive dysfunction in 55%, seizure disorder in 51%, acute confusional disorder in 35%, peripheral nervous system impairment in 15%, psychosis in 12%, and stroke in 12%.

Somewhat strangely, although pediatric SLE patients were found to present more frequent renal disease and encephalopathy than adults [51], long-term survival for pediatric patients with NPSLE was as high as 97% [52].

Of note, the presence of antiphospholipid antibodies was seen in 70% of children as compared with 25–30% in adult SLE patients in the study by Harel et al. [53]. However, their association with NPSLE was unremarkable except for CVD [53]. Although neurocognitive deficits have been found in 55–59% of pediatric SLE patients [46, 54], the true prevalence rates and impacts on academic performance as well as on quality of life status remain to be elucidated [55].

1.6 Steroid Induced Psychosis

It is often difficult to determine whether psychiatric manifestations are caused by SLE itself (NPSLE) or induced by steroids (steroid psychosis), especially when the patients present the manifestations during the course of SLE. Steroid psychosis is defined as psychiatric syndrome that newly appears after the introduction or increase of steroids in patients with SLE. The incidence of steroid psychosis appears to be much lower than expected. Thus, it has been reported that approximately 5.4% (28/520) presented steroid psychosis [56]. Similar incidence rate (5%) of steroid psychosis has been reported in the other study [57].

It should be noted that about 50% of patients with diffuse NPSLE presented apparent manifestations after the induction or increase of steroid. Therefore, surrogate markers for diffuse NPSLE, such as CSF IL-6 [58], would be important and useful for the correct diagnosis. It is also notable that ACS is much more frequently observed in diffuse NPSLE than in steroid psychosis, whereas mood

Table 1.4 Comparison of neuropsychiatric manifestations between lupus psychosis and steroid psychosis

Neuropsychiatric manifestations	Lupus psychosis (NPSLE)[a] (n = 37)	Steroid-induced psychosis (n = 25)
Acute confusional state	22	0
Psychosis	4	3
Anxiety disorder	2	2
Mood disorder	7	19
Cognitive dysfunction	2	1

[a]One patient showed anxiety disorder and mood disorder, one patient presented acute confusional state and psychosis and one patient had acute confusional state with anxiety disorder and mood disorder

disorder is much more common in steroid psychosis in our series of patients (Table 1.4).

1.7 Mortality

It has been reported that patients with SLE have a 3 fold increased risk of mortality with standardized mortality ratio (SMR) [59, 60], the ratio between the observed number of deaths of patients and the expected number of deaths in the general population adjusted for age, sex and time of diagnosis. SMR increases to 6 in lupus nephritis, and to 9.5 (95% CI 6.7–13.5) in NPSLE [60]. Thus, in NPSLE the five-year survival estimate and the 10-year survival estimate were 0.85 and 0.76, respectively [60]. As can be expected, hazard ratios (HRs) were highest in patients with acute confusional state (HR 3.4). We have recently demonstrated that the presence of abnormalities on MRI scans significantly increased the mortality of patients with ACS (5-year survival 0.55 compared with 1.9 in patients without MRI abnormalities) [61]. In patients with NPSLE, most common causes of death were infection and NPSLE itself [60].

Of interest, a decreased mortality risk was seen in patients with antiplatelet therapy (HR 0.22) [60]. It is suggested that such effect of anti-platelet therapy might be due to its action on CVD or atherosclerosis. However, prospective studies need to be done to draw any definite conclusion.

1.8 Emerging Problems from the Epidemiological Studies

As mentioned above, the most critical issue on the 1999 ACR nomenclature and definition [3] is that a minor and milder manifestations are mixed up with a serious manifestation within each category, especially headache, cognitive dysfunction and mood disorders [3]. For example, the prevalence of all headache types, particularly that of tension-type headache and migraine, does not differ between SLE patients and the general population [62]. Thus, it remains unclear which type of headache is caused by immunological processes related with SLE and requires immunosuppressive therapy. In this regard, it has been found that headache due to intracranial hypertension and intractable non-specific headache, but not migraine, are characterized by the inflammatory profile in CSF, such as the elevation of IL-6, IL-8, and IP-10 [63], providing rationale for immunosuppressive therapy.

The ACR committee also proposed a standard battery of neuropsychologic tests to detect cognitive dysfunction [3]. However, such a test would be sometimes confusing, resulting in inclusion of patients with subclinical cognitive dysfunction who do not require immunosuppressive therapy. It is therefore likely that the prevalence of cognitive dysfunction in retrospective studies might reflect the real prevalence of patients who require immunosuppressive therapy. Similarly, as to mood disorders, the retrospective studies might provide the more likely prevalence of patients requiring immunosuppressive therapy. Some testing is necessary to determine whether the patients require the immunosuppressive therapy instead of detecting very subtle changes in cognitive function. In this regard, CSF IL-6 has been shown to be useful to make a differential diagnosis of patients with diffuse NPSLE, including cognitive dysfunction and mood disorders, from those with psychiatric manifestations not requiring immunosuppressive therapy [64].

In conclusion, although the prevalence of headache, cognitive dysfunction and mood disorders is relatively high in the prospective studies, the percentage of patients who required immunosuppressive therapy is very limited, possibly accounting for the lower prevalence of these manifestations in the retrospective studies. It should be emphasized that the establishment of surrogate markers to detect patients requiring immunosuppressive therapy.

1.9 Summary

A variety of neuropsychiatric manifestations occur in patients with SLE. The ACR developed standardized nomenclature and case definitions for NPSLE in 1999, which has enabled the epidemiological studies to be performed on an equal basis. One of the problems in the 1999 ACR classification is the inclusion of minor and milder and less specific manifestations, including headache, mild cognitive dysfunction and mood disorders, resulting in some confusion in the decision of immunosuppressive therapy. Another critical point of the ACR classification is the lack of attention to a number of other neurological manifestation, especially NMOSD, posterior reversible encephalopathy syndrome and reversible focal neurological deficits mimicking CVD. The overall prevalence of NP SLE was estimated to be 56.3%, predominantly affecting CNS (93.1%) rather than PNS (6.9%). Hispanics, African descendants and Asians were found to develop NPSLE more frequently as well as severely than Caucasians. NPSLE usually occur early in the course of SLE. It should be kept in mind that steroid psychosis is sometimes a differential diagnosis, but not necessarily an exclusion, of lupus psychosis. NPSLE increased the mortality rate with SMR 9.5. It should be emphasized that the establishment of surrogate markers to detect patients with headache, cognitive dysfunction and mood disorders requiring immunosuppressive therapy is important for a better management of patients with NPSLE.

References

1. Gibson T, Myers AR. Nervous system involvement in systemic lupus erythematosus. Ann Rheum Dis. 1975;35:398–406.
2. Harris EN, Hughes GR. Cerebral disease in systemic lupus erythematosus. Springer Semin Immunopathol. 1985;8:251–66.
3. ACR Ad Hoc Committee on Neuropsychiatric Lupus Nomenclature. The American College of Rheumatology nomenclature and case definitions for neuropsychiatric lupus syndromes. Arthritis Rheum. 1999;42:599–608.
4. Cohen AS, Canoso JJ. Criteria for the classification of systemic lupus erythematosus-status 1972. Arthritis Rheum. 1972;15:540–3.
5. Tan EM, et al. The 1982 revised criteria for the classification of systemic lupus erythematosus. Arthritis Rheum. 1982;25:1271–7.
6. Grigor R, et al. Systemic lupus erythematosus. Ann Rheum Dis. 1978;37:121–8.
7. Kremer JM, et al. Non-organic, non-psychotic psychopathology (NONPP) in patients with systemic lupus erythematosus. Semin Arthritis Rheum. 1981;11:182–9.
8. Zvaifler NJ, Bluestein HG. The pathogenesis of central nervous system manifestations of systemic lupus erythematosus. Arthritis Rheum. 1982;25:862–6.
9. Arinuma Y, et al. Association of cerebrospinal fluid anti-NR2 glutamate receptor antibodies with diffuse neuropsychiatric systemic lupus erythematosus. Arthritis Rheum. 2008;58:1130–5.
10. Bertsias GK, Iet a. EULAR recommendations for the management of systemic lupus erythematosus with neuropsychiatric manifestations: report of a task force of the EULAR standing committee for clinical affairs. Ann Rheum Dis. 2010;69:2074–82.
11. Steup-Beekman GM, et al. Neuropsychiatric manifestations in patients with systemic lupus erythematosus: epidemiology and radiology pointing to an immune-mediated cause. Ann Rheum Dis. 2013;72(Suppl 2):ii76–9.
12. Govoni M, et al. The diagnosis and clinical management of the neuropsychiatric manifestations of lupus. J Autoimmun. 2016;74:41–72.
13. Ainiala H, et al. Validity of the new American College of Rheumatology criteria for neuropsychiatric lupus syndromes: a population-based evaluation. Arthritis Rheum. 2001;45:419–23.
14. Brey RL, et al. Neuropsychiatric syndromes in lupus: prevalence using standardized definitions. Neurology. 2002;58:1214–20.
15. Hanly JG, et al. Systemic lupus international collaborating clinics, neuropsychiatric events at the time of diagnosis of systemic lupus erythematosus: an international inception cohort study. Arthritis Rheum. 2007;56:265–73.
16. Bertsias GK, Boumpas DT. Pathogenesis, diagnosis and management of neuropsychiatric SLE manifestations. Nat Rev Rheumatol. 2010;6:358–67.
17. Sanna G, et al. Neuropsychiatric manifestations in systemic lupus erythematosus: prevalence and association with antiphospholipid antibodies. J Rheumatol. 2003;30:985–92.
18. Cervera R, et al. European working party on systemic lupus erythematosus, lessons from the "euro-lupus cohort". Ann Med Interne. 2002;153:530–6.
19. Unterman A, et al. Neuropsychiatric syndromes in systemic lupus erythematosus: a meta-analysis. Semin Arthritis Rheum. 2011;41:1–11.

20. Borowoy AM, et al. Neuropsychiatric lupus: the prevalence and autoantibody associations depend on the definition: results from the 1000 faces of lupus cohort. Semin Arthritis Rheum. 2012;42:179–85.
21. Amur S, et al. Sex differences and genomics in autoimmune diseases. J Autoimmun. 2012;38:J254–65.
22. Thumboo J, et al. A comparative study of the clinical manifestations of systemic lupus erythematosus in Caucasians in Rochester, Minnesota, and Chinese in Singapore, from 1980 to 1992. Arthritis Rheum. 2001;45:494–500.
23. Uribe AG, Alarcon GS. Ethnic disparities in patients with systemic lupus erythematosus. Curr Rheumatol Rep. 2003;5:364–9.
24. Fernandez M, et al., LUMINA Study Group. A multiethnic, multicenter cohort of patients with systemic lupus erythematosus (SLE) as a model for the study of ethnic disparities in SLE. Arthritis Rheum 2007; 57:576–584.
25. Gonzalez LA, et al. Impact of race and ethnicity in the course and outcome of systemic lupus erythematosus. Rheum Dis Clin N Am. 2014;40:433–54.
26. Rivest C, et al. Association between clinical factors, socioeconomic status, and organ damage in recent onset systemic lupus erythematosus. J Rheumatol. 2000;27:680–4.
27. Govoni M, et al., Italian Society of Rheumatology. Factors and comorbidities associated with first neuropsychiatric event in systemic lupus erythematosus: does a risk profile exist? A large multicentre retrospective cross-sectional study on 959 Italian patients. Rheumatology. 2012; 51:157–168.
28. Andrade RM, et al., LUMINA Study Group. Seizures in patients with systemic lupus erythematosus: data from LUMINA, a multiethnic cohort (LUMINA LIV). Ann Rheum Dis 2008; 67:829–834,
29. Mikdashi J, Handwerger B. Predictors of neuropsychiatric damage in systemic lupus erythematosus: data from the Maryland lupus cohort. Rheumatology. 2004;43:1555–60.
30. Tomietto P, et al. General and specific factors associated with severity of cognitive impairment in systemic lupus erythematosus. Arthritis Rheum. 2007;57:1461–72.
31. Hanly JG, et al. Seizure disorders in systemic lupus erythematosus results from an international, prospective, inception cohort study. Ann Rheum Dis. 2012;71:1502–9.
32. Bujan S, et al. Contribution of the initial features of systemic lupus erythematosus to the clinical evolution and survival of a cohort of Mediterranean patients. Ann Rheum Dis. 2003;62:859–65.
33. Mikdashi J, et al. Factors at diagnosis predict subsequent occurrence of seizures in systemic lupus erythematosus. Neurology. 2005;64:2102–7.
34. Appenzeller S, et al. Epileptic seizures in systemic lupus erythematosus. Neurology. 2004;63:1808–12.
35. Mok MY, et al. Antiphospholipid antibody profiles and their clinical associations in Chinese patients with systemic lupus erythematosus. J Rheumatol. 2005;32:622–8.
36. McLaurin EY, et al. Predictors of cognitive dysfunction in patients with systemic lupus erythematosus. Neurology. 2005;64:297–303.
37. Hanly JG, et al. Autoantibodies as biomarkers for the prediction of neuropsychiatric events in systemic lupus erythematosus. Ann Rheum Dis. 2011;70:1726–32.
38. Narvaez J, et al. Rituximab therapy in refractory neuropsychiatric lupus: current clinical evidence. Semin Arthritis Rheum. 2011;41:364–72.
39. Hirohata S, et al. Blood-brain barrier damages and intrathecal synthesis of anti-N-methyl-D-aspartate receptor NR2 antibodies in diffuse psychiatric/neuropsychological syndromes in systemic lupus erythematosus. Arthritis Res Ther. 2014;16:R77.
40. Gladman DD, et al. Systemic lupus erythematosus disease activity index 2000. J Rheumatol. 2002;29:288–91.
41. Isenberg DA, et al. 2004 BILAG, development and initial validation of an updated version of the British isles lupus assessment Group's disease activity index for patients with systemic lupus erythematosus. Rheumatologu. 2005;44:902–6.

42. Brandt KD, Lessell S. Migrainous phenomena in systemic lupus erythematosus. Arthritis Rheum. 1978;21:7–16.
43. Toubi E, et al. Association of antiphospholipid antibodies with central nervous system disease in systemic lupus erythematosus. Am J Med. 1995;99:397–401.
44. Kimura M, et al. Reversible focal neurological deficits in systemic lupus erythematosus: report of 2 cases and review of the literature. J Neurol Sci. 2008;272:71–6.
45. Birnbaum J, et al. Distinct subtypes of myelitis in systemic lupus erythematosus. Arthritis Rheum. 2009;60:3378–87.
46. Sibbitt WL Jr, et al. The incidence and prevalence of neuropsychiatric syndromes in pediatric onset systemic lupus erythematosus. J Rheumatol. 2002;29:1536–42.
47. Muscal E, Myones BL. The role of autoantibodies in pediatric neuropsychiatric systemic lupus erythematosus. Autoimmun Rev. 2007;6:215–7.
48. Brunner HI. Risk factors for damage in childhood-onset systemic lupuserythematosus: cumulative disease activity and medication use predict disease damage. Arthritis Rheum. 2002;46:436–44.
49. Brunner HI, et al. Difference in disease features between childhood-onsetand adult-onset systemic lupus erythematosus. Arthritis Rheum. 2008;58:556–62.
50. Tucker LB, et al. Adolescent onset of lupus results in more aggressive disease and worse outcomes: results of a nested matched case-control study within LUMINA, a multiethnic US cohort (LUMINA LVII). Lupus. 2008;17:314–22.
51. Hoffman IEA, et al. Juvenile-onset systemic lupus erythemotosis: different clinical and serological pattern than adult-onset systemic lupus erythematosus. Ann Rheum Dis. 2009;68:412–5.
52. Benseler SM, Silverman ED. Neuropsychiatric involvement in pediatric systemic lupus erythematosus. Lupus. 2007;16:564–71.
53. Harel L, et al. Neuropsychiatric manifestations in pediatric systemic lupus erythematosus and association with antiphospholipid antibodies. J Rheumatol. 2006;33:1873–7.
54. Brunner HI, et al. Initial validation of the pediatric automated neuropsychological assessment metrics for childhood-onset systemic lupus erythematosus. Arthritis Rheum. 2007;57:1174–82.
55. Muscal E, Brey RL. Neurological manifestations of systemic lupus erythematosus in children and adults. Neurol Clin. 2010;28:61–73.
56. Appenzeller S, et al. Acute psychosis in systemic lupus erythematosus. Rheumatol Int. 2008;28:237–43.
57. Chau SY, Mok CC. Factors predictive of corticosteroid psychosis in patients with systemic lupus erythematosus. Neurology. 2003;61:104–7.
58. Nishimura K, et al. Blood-brain barrier damage as a risk factor for corticosteroid-induced psychiatric disorders in systemic lupus erythematosus. Psychoneuroendocrinology. 2008;33:395–403.
59. Yurkovich M, et al. Overall and cause-specific mortality in patients with systemic lupus erythematosus: a meta-analysis of observational studies. Arthritis Care Res (Hoboken). 2014;66:608–16.
60. Zirkzee EJ, et al. Mortality in neuropsychiatric systemic lupus erythematosus (NPSLE). Lupus. 2014;23:31–8.
61. Abe G, et al. Brain MRI in patients with acute confusional state of diffuse psychiatric/neuropsychological syndromes in systemic lupus erythematosus. Mod Rheumatol. 2017;27:278–83.
62. Unterman A, et al. Neuropsychiatric syndromes in systemic lupus erythematosus: a meta-analysis. Semin Arthritis Rheum. 2011;41:1–11.
63. Fragoso-Loyo H, et al. Inflammatory profile in cerebrospinal fluid of patients with headache as a manifestation of neuropsychiatric systemic lupus erythematosus. Rheumatology (Oxford). 2013;52:2218–22.
64. Hirohata S, et al. Accuracy of cerebrospinal fluid IL-6 testing for diagnosis of lupus psychosis. A multicenter retrospective study. Clin Rheumatol. 2009;28:1319–23.

Chapter 2
Genetics

Naoyuki Tsuchiya

Abstract Extensive studies revealed more than 70 strong candidate regions for susceptibility genes to systemic lupus erythematosus (SLE), and efforts to identify the causative variants in each candidate region are under way. The list of candidate genes points to the crucial pathways that play a role in the development of SLE, such as HLA and immune system signaling, upregulated type I interferon and nucleic acids response, and defective clearance of dying cells. Among these pathways, type I interferon pathway may be particularly relevant to neuropsychiatric SLE (NPSLE), because Aicardi-Goutières syndrome (AGS), a group of single gene diseases with enhanced type I IFN response and exhibits severe central nervous system symptoms, has some similarities with SLE. In fact, variants in some of the genes responsible for AGS are also reported in familial and sporadic patients with SLE. On the other hand, the efforts to identify NPSLE associated genes using case-case association analysis have not been very successful thus far. In the future, large-scale case-case association analysis, not limited to the genes associated with overall SLE, may be necessary in order to identify variants associated with clinical subphenotypes including neuropsychiatric manifestations.

Keywords Genetics · Systemic Lupus Erythematosus · Neuropsychiatric · Type I interferon · *TREX1*

2.1 Introduction

Although the etiology of systemic lupus erythematosus (SLE) remains unclear, several lines of evidence from genetic epidemiology strongly support the hypothesis that the majority of the patients with SLE are caused by a combination of multiple genetic as well as environmental factors. Although SLE is not inherited in a mendelian

---

N. Tsuchiya (✉)
Molecular and Genetic Epidemiology Laboratory, Faculty of Medicine,
University of Tsukuba, Tsukuba, Japan
e-mail: tsuchiya-tky@umin.net

© Springer International Publishing AG, part of Springer Nature 2018
S. Hirohata (ed.), *Neuropsychiatric Systemic Lupus Erythematosus*,
https://doi.org/10.1007/978-3-319-76496-2_2

fashion except for a small number of patients, siblings of SLE patients have 8~29 fold higher risk for developing SLE when compared with general population [1]. Concordance between monozygotic twins is much higher than between dizygotic twins (24% versus 2%) [2]. In addition, the prevalence of SLE is higher in populations of African and Asian ancestry as compared with European-ancestry populations [3].

Efforts to detect genetic variants associated with SLE started in the 1970's, using candidate gene approach. Since the 2000's, genome-wide association studies (GWAS) designed to genotype millions of single nucleotide polymorphisms (SNPs) across the genome identified a number of strong candidate regions across the genome that are associated with susceptibility to or protection against SLE. Subsequently, new methodologies such as microarrays to genotype SNPs in the immune system genes across the genome (Immunochip) [4, 5] and imputation of GWAS [6] to combine the studies on multiple populations increased the number of convincing candidate regions (reviewed in [7]). Furthermore, attempts to pinpoint the functional variants which can explain the genetic association of each region ("causative" variants) using next generation sequencing have started [8].

It has been proposed that SLE might be actually composed of multiple subgroups defined by different sets of clinical characteristics [9]. If this is the case, some genetic variants may be associated with specific clinical characteristics, although efforts to identify such variants have not been very fruitful thus far.

In this chapter, I will review the current understanding of genetics of overall SLE. I will also discuss the attempts to find the gene variants specifically associated with neuropsychiatric SLE (NPSLE), although this area of research still appears to be an open field.

2.2 Genetics of Overall SLE

Table 2.1 lists more than 70 chromosomal regions encompassing either the variants with genome-wide significance ($P < 5 \times 10^{-8}$), replicated candidate gene variants, or rare variants reported to be causative in familial or possibly mendelian forms of SLE. The plausible candidate genes in each region are also listed; however, only in a small number of these regions the causative variants driving the genetic association have been convincingly identified, along with their molecular mechanisms. Nevertheless. the list of the candidate genes points to several molecular pathways which probably play a crucial role in the development of SLE. I will discuss some of such pathways.

2.2.1 Major Histocompatibility Complex (MHC) Region

In the Caucasian populations, the haplotypes carrying *HLA-DRB1*03:01, DRB1*15:01* and *DRB1*08:01* have been shown to be associated with SLE [10]. *DRB1*03:01* is encoded by a haplotype *A*01:01-B*08:01-C*07:01-DRB1*03:01-DQA1*05:01-DQB1*02:01* (*DRB1*03:01* haplotype), which is the most prevalent

Table 2.1 Plausible candidate susceptibility genes to systemic lupus erythematosus in 2017

Chromosome	Plausible candidate genes	Study design	Suggested molecular pathways of association
1	C1Q	Familial SLE	Clearance of dead cells and immune complexes
	PTPN22	Candidate gene, GWAS	Inhibitory signaling in T cells
	FCGR2A, 3A, 3B	Candidate gene, GWAS	Immune complex clearance
	FCGR2B	Candidate gene	Inhibitory signaling in B cells, dendritic cells and monocytes
	TNFSF4 (OX40L)	Candidate gene, GWAS	T cell activation
	NCF2, SMG7	GWAS	NADPH oxidase subunit (NCF2), nonsense mediated mRNA decay (SMG7)
	PTPRC (CD45)	GWAS	Signaling in T cells and B cells
	IKBKE	Candidate gene, GWAS	NF-κB activation
	IL10	Candidate gene, GWAS	B cell activation, Th1/Th2 balance
	LYST	GWAS	Lysosomal trafficking regulation
2	LBH	GWAS	
	SPRED2	GWAS	
	IFIH1(MDA5)	Candidate gene, GWAS	Cytoplasmic RNA sensor, type I interferon(IFN) induction
	STAT4	Candidate gene, GWAS	IL-12 signaling, Th1 differentiation
	IKZF2	GWAS	Lymphocyte differentiation
	RASGRP3	GWAS	Regulation of TLR signal
3	ABHD6, PXK	GWAS	
	IL12A	GWAS	Th1 differentiation
	TREX1	Rare variant	Cytoplasmic DNA degradation
	DNASE1L3	Familial SLE	Degradation of DNA in apoptotic microparticles
	TMEM39A	GWAS, Immunochip	
4	BANK1	GWAS	B cell signaling
	DGKQ	Immunochip	
5	TCF7, SKP1	GWAS	Transcription factor in T cells (TCF7), ubiquitination (SKP1)
	TNIP1	GWAS	Regulation of NF-κB activation
	MIR146A	GWAS	Anti-inflammatory effect
	IL12B	Immunochip	IL-12 and IL-23 signals involved in T cell activation
	TERT	Immunochip	Teromerase reverse transcriptase

(continued)

Table 2.1 (continued)

Chromosome	Plausible candidate genes	Study design	Suggested molecular pathways of association
6	*ATXN1*	GWAS	
	MHC region	Candidate gene, GWAS	Antigen presentation, immunoregulation (*HLA* genes), clearance of apoptotic cells and immune complexes (*C4, C2*)
	DEF6	Immunochip	Regulation of IRF4
	UHRF1BP1	GWAS	
	BACH2	GWAS	Regulation of antibody production in B cells
	PRDM1, ATG5	GWAS	IFNβ suppression, B cell differentiation (*PRDM1*), autophagy (*ATG5*)
	TNFAIP3	GWAS	Regulation of NF-κB activation
	LRRC16A(CARMIL1)	Immunochip	
	SLC17A4	Immunochip	Sodium/phosphate cotransporter in the intestinal mucosa
7	*JAZF1*	GWAS	Transcriptional repressor
	IKZF1	GWAS	Lymphocyte differentiation
	GTF2IRD1, GTF2I, NCF1	Immunochip	Transcription factor (*GTF2IRD1, GTF2I*), NADPH oxidase subunit (*NCF1*)
	IRF5	Candidate gene, GWAS	Induction of type I IFN, IFN stimulated genes and proinflammatry cytokines
8	*BLK*	GWAS	B cell signaling
	FAM86B3P	Immunochip	
	PLAT	Immunochip	Tissue type plasminogen activator
9	*JAK2*	GWAS	Cytokine receptor signaling
10	*WDFY4*	GWAS	
	ARID5B	GWAS	Transcription activator
11	*IRF7*	Candidate gene, GWAS	Induction of type I IFN and IFN stimulated genes
	CD44	Linkage study, GWAS	Lymphocyte activation, homing, apoptosis
	RNASEH2C	GWAS	Degradation of RNA in RNA-DNA hybrids
	DHCR7, NADSYN1	GWAS	
	ETS1, FLI1	GWAS	Transcription factors
	PCNX3	Immunochip	Notch signaling
12	*SH2B3*	GWAS	Regulatory adaptor
	SLC15A4	GWAS	Involved in TLR7, TLR9 induced type I IFN production
	C1R, C1S	Familial SLE	Clearance of dead cells and immune complexes

(continued)

Table 2.1 (continued)

Chromosome	Plausible candidate genes	Study design	Suggested molecular pathways of association
13	TNFSF13B (BAFF)	GWAS	B cell activation and differentiation
14	RAD51B	GWAS	Apoptosis
15	CSK	GWAS	Regulation of B cell signaling
	RASGRP1	Immunochip	T cell and B cell differentiation
16	CIITA, SOCS1	Candidate gene, GWAS	Regulation of MHC class II expression (CIITA), regulation of cytokine signaling (SOCS1)
	ITGAM	GWAS	Complement-mediated phagocytosis, cell adhesion
	ZFP90 (FIK)	GWAS	Transcription factor
	IRF8	GWAS	Induction of type I IFN and IFN-stimulated genes
	DNASE1	Rare variant	Degradation of apoptotic cell-derived DNA
17	PLD2	GWAS	Encodes phospholipase D2 involved in signal transduction
	IKZF3	GWAS	Lymphocyte differentiation
	CD226	Immunochip	Adhesion of platelets to endothelial cells. NK cells and T cells functions
	GRB2	Immunochip	Adaptor protein
19	TYK2	Candidate gene, GWAS	Type I IFN signal transduction
	SIGLEC6	Immunochip	Cell-cell adhesion involving B cells or placenta
22	UBE2L3	GWAS	Ubiquitin conjugating enzyme, possibly involved in NF-kB activation
	SYNGR1	Immunochip	Neuronal synaptic transmissiom
X	TLR7, TLR8	Candidate gene	Recognize single strand RNA and activate type I IFN pathway
	CXorf21	GWAS	
	IRAK1	Candidate gene, GWAS	IL-1 signal transduction
	MECP2	Candidate gene, GWAS	Transcription regulation by binding to methylated DNA

The regions encompassing either the variants with genome-wide significance ($P < 5 \times 10^{-8}$), replicated candidate gene variants, or rare variants reported to be causative in familial or possibly mendelian form of SLE are listed
IFN interferon

haplotype among the northern European populations and is associated with multiple autoimmune diseases.

In the Asians, the frequency of *DRB1*03:01* in the general population is low, and *DRB1*15:01* has been shown to be associated with SLE in the East Asian populations [11], and possibly *DRB1*15:02* in the Southeast Asian populations [12, 13]. Recent findings from Japan indicated protective association of *DRB1*13:02* and *DRB1*14:03* with SLE [11]. Interestingly, *DRB1*13:02* is associated with protection against rheumatoid arthritis [14], antineutrophil cytoplasmic antibody (ANCA) associated vasculitis (AAV) [15], systemic sclerosis [16] and polymyositis/ dermatomyositis [17] at least in the Japanese population. Thus, *DRB1*13:02* is a shared protective allele against multiple autoimmune rheumatic diseases [18].

In the *MHC* region, potentially relevant genes to SLE such as *C4* and *C2* are encoded, and wide-range linkage disequilibrium (LD) is observed. Thus, it is possible that variants on loci other than *HLA* might play a primary role, or at least have an independent contribution. Independent contribution of *DRB1-DQA1*, *DPB1*, *MSH5*, *HLA-B* and *HLA-G* variants has already been reported [19, 20]. In addition, *C4* gene has highly complicated copy number variations, and some studies suggested independent genetic contribution of *C4* genes [21].

Until recently, it was generally believed that the mechanism of association between *HLA* and diseases may be explained by allele-specific presentation of antigenic peptides to T cells. Although such pathogenic antigen peptides have not been identified, recent studies reported that the genetic contribution of the *MHC* region can be ascribed to specific amino acid sequences which constitute antigenic peptide binding sites [22]. On the other hand, deep sequencing of *HLA-D* region identified regulatory SNPs in the intergenic regions with independent genetic effects, and the risk genotypes were associated with increased expression of HLA-DR and DQ molecules. These functional regulatory SNPs are in LD with *HLA-DR* or *DQ* alleles, suggesting that both the expression levels and amino acid sequences of HLA class II peptide binding grooves play a role in the development of SLE in combination [8].

2.2.2 Type I Interferon Pathway and Nucleic Acid Response Genes

Genes induced by type I interferon (IFN) are strikingly upregulated in the peripheral blood leukocytes from SLE ("type I IFN signature") [23, 24]. In addition, variants in a number of type I IFN induction pathway genes including *IRF5* [25, 26], *TLR7* [27, 28], *IFIH1, IRF8 and TYK2* [29] have been associated with SLE. In general, the risk alleles are associated with enhanced induction of type I IFN and IFN stimulated genes [30]. These findings suggest that individuals with predisposition to enhanced response of type I IFN are associated with SLE.

The major triggers of type I IFN response are nucleic acids, most typically viral DNA and RNA. Recently, Crow referred a group of autoinflammatory or antoimmune disorders caused by single gene mutations in which upregulation of type I IFN plays a critical role in the pathogenesis "type I interferonopathies" [31]. The most typical type I interferonopathy is Aicardi-Goutières syndrome (AGS), characterized by early-onset neurological disorders including basal ganglia calcification, brain atrophy, increase in the lymphocytes and IFNα level in the cerebrospinal fluid. Although AGS is by no means identical to SLE, skin lesion similar to SLE (chilblain lupus) is frequently observed, and a recent study reported that a variety of autoantibodies, including antinuclear antibodies and brain-reactive antibodies, are frequently detected in the patients with AGS [32]. In fact, some patients meet the American College of Rheumatology classification criteria for SLE [31, 33] . Elevated level of IFNα in the cerebrospinal fluid was also described in lupus psychosis [34].

Mutations that cause AGS are located most frequently in *TREX1*, which degrades single-stranded and double-stranded DNA (ssDNA and dsDNA), *RNASEH2A*,

RNASEH2B and *RNASEH2C*, which are the subunits of RNase H2 complex that degrades RNA in RNA-DNA hybrids, and *SAMHD1*, which hydrolyses deoxynucleotides and may also degrade RNA [31]. Loss-of-function mutations of these genes may cause abnormal accumulation of nucleic acids and trigger type I IFN response. In addition, gain-of-function mutation of *IFIH1* has also been reported in AGS [35].

Notably, variants of *TREX1*, *RNASEH2A/2B/2C*, *SAMHD1* and *IFIH1* have been associated with SLE either in a monogenic or polygenic fashion [29, 36–39]. Thus, the similarity between AGS and SLE, especially NPSLE, may provide a clue to elucidate the pathogenesis of NPSLE.

2.2.3 Defective Clearance of Dying Cell Nucleic Acids

A major source of extracellular nucleic acids is from dying or dead cells. Impaired clearance of dead or dying cells has been shown in SLE [40]. Complete deficiency of early complement components such as C1q, C1r, C1s, C4 and C2 are very strongly associated with SLE [41], and the mechanism is thought to involve defective clearance of dead or dying cells. This leads to leakage of nucleosomes to extracellular fluid, which are recognized by immune system, and results in production of autoantibodies against dsDNA, ssDNA, RNA-protein complexes and histones.

Defective clearance of apoptotic cells also results in release of microparticles containing chromatin. Microparticle chromatin is usually digested by *DNASE1L3*, but loss-of-function mutation of this gene has been reported in familial SLE patients [36, 42].

Furthermore, neutrophils undergo NETosis when stimulated for example by bacterial infection, and extrude DNA, histone and various proteins which are targets of autoantibodies such as myeloperoxidase, proteinase 3 and citrullinated proteins. These molecules form neutrophil extracellular traps (NETs). Normally, NETs are degraded by DNase 1, the deficiency of which has been reported in SLE [43]. Thus, it is possible that accumulation of extracellular DNA or RNA may trigger type I IFN response via *TLR3*, *TLR7* and *TLR9*.

2.2.4 Signaling Molecules in Immune System Cells

A substantial number of candidate susceptibility genes to SLE are involved in signaling in the immune system cells. Many of them are regulators of NF-κB pathway, including *TNFAIP3* which encodes A20 [44], *TNIP1* which encodes ABIN1 [45], *IKBKE* [46] and *UBE2L3* [47]. In general, the risk alleles of these genes are suggested to be associated with activation, and the protective allele with regulation, of NF-kB activation.

B cell signaling molecules including *BLK*, which appears to regulate B cell receptor signaling [48–51], and T cell signaling molecules including *STAT4*, which is involved in IL-12 signaling to induce Th1 differentiation [51, 52], are also strongly associated with SLE, underscoring the crucial role of adaptive immunity.

Recently, studies on Asian populations revealed that the region on chromosome 7 encompassing transcription factor genes *GTF2IRD1* and *GTF2I* is strikingly associated with SLE, even more strongly than *MHC* region, in the Asian populations [5]. This chromosomal region has a highly complicated structure, and also includes *NCF1*, a subunit of NADPH oxidase, and one study reported that the association signal is primarily ascribed to *NCF1* [53]. Further studies will be necessary to determine the molecular mechanisms of association of this region.

2.3 Genetics of Neuropsychiatric SLE

In contrast to overall SLE, variants specifically associated with clinical subsets remain largely undefined. In order to detect variants associated with NPSLE, the patients with NPSLE should be compared with SLE without NP symptoms. Because NPSLE accounts for a rather small proportion of SLE, most association studies reported thus far analyzed small sample size, and convincing association has not been reported. However, *TREX1*, whose mutations constitute the most frequent cause of AGS which exhibits serious neurological complications, has been intensively examined for rare as well as common variants in several studies. Here I will review some reports which implicated the role of *TREX1* rare variants as well as common SNPs in NPSLE. Reports on familial chilblain lupus with cerebral vasculitis were excluded. Studies suggesting association of other genes are also briefly discussed.

2.3.1 TREX1

De Vries et al. directly sequenced genomic DNA from 60 NPSLE patients for exonic *TREX1* mutations, and identified a novel heterozygous missense mutation (R128H) in one NPSLE patient. Brain MRI showed generalized atrophy, extensive symmetric cerebral white matter hyperintensities and cerebellar infarcts without evidence for recent ischemia [38].

Ellyard et al. performed exome sequencing in a 4-year-old girl with cerebral lupus and identified a rare, homozygous mutation R97H in *TREX1* that was predicted to be highly deleterious. This product had a 20-fold reduction in exonuclease activity and was associated with an elevated IFNα signature [37].

Finally, Fredi et al. sequenced *TREX1* gene in 51 SLE patients, and identified a novel heterozygous variant (D130N) in one patient and in none of 150 controls. Interestingly, when the eight patients with NP manifestations were compared with 43 patients without NP manifestations, a significantly higher minor allele frequency of SNP rs11797 encoding a synonymous substitution was found in NPSLE (12/16 [75%] vs 28/86 [32.5%], P = 0.002, odds ratio [OR] 6.42) [54], suggesting that not only rare variants, but common SNPs in *TREX1* may also be associated with NPSLE. However, the number of patients with NPSLE was very small (n = 8), and this interesting observation requires replication.

2.3.2 Other Candidate Genes

A meta-analysis on NPSLE association studies on *IL1B*, *IL1RN*, *IL6*, *IL10*, *ITGAM*, *MBL2*, *TNF*, *FCGR2A/3A/3B*, *XRRCC1* and *VDR* genes was reported in 2015 [55]. Among these genes, Fcγ receptor genes reached nominal significance. *FCGR3A*-158F/F genotype was associated with NPSLE (OR 1.887, P = 0.026). *FCGR3B*-NA1/NA2 was increased as compared with NA1/NA1 in NPSLE (OR 2.141, P = 0.032), *FCGR2A*-131H/H was increased in NPSLE as compared with 131R/R genotype (OR 3.113, P = 0.048). *FCGR3A*-158F (also referred to as 176F) has weaker affinity to IgG1 and IgG3 compared with 158V, and *FCGR3B*-NA2 has lower capacity of phagocytosis. Both alleles have been reported to be associated with overall SLE in some studies. On the other hand, *FCGR2A*-131H binds IgG2, while 131R does not bind IgG2, and association with overall SLE has been reported for 131R rather than 131H. Therefore, the interpretation of association of these *FCGR* alleles seems to be difficult, and because the sample size is rather small, further studies are required.

Striking association with NPSLE was reported in the SNPs in *CD244*, a SLAM family member, in a Japanese population. The frequency of the rs6682654C allele and rs3766379T allele was significantly increased in NPSLE (P = 1.63×10^{-7}, OR 3.47 and P = 2.55×10^{-7}, OR 3.15), respectively [56]. This finding awaits further replication study.

APOE E4 has been reported to be significantly different between NPSLE and non-NPSLE groups (P = 3.26×10^{-5}, OR 6.81, calculated by the author based on the published data) from Slovakia [57]. Other studies showed nominal associations of genes such as *FCGR2B*, *IL10*, *PXK*, *BANK1*, *IL21*, *IRF5*, *XKR6* [58], *VEGF* [59], *TRPC6* [60], *DEFB* [61], *ESR1* [62], *ITGAM* [63], *XRCC1*, *XRCC3*, *XRCC4* [64], *BDNF* [65] and *RANTES* [66]. Overall, most of these studies reported modest associations in rather small sample size. Further replication studies will be necessary.

2.4 Summary

Dozens of strong candidates for the susceptibility genes to sporadic and familial SLE have been reported based on candidate gene studies, genome-wide studies and sequencing studies. These studies lead to identification of critical pathways such as type I IFN and nucleic acid response pathway, dying cell clearance pathway and immune system signaling pathways, and opened a number of new fields for research. On the other hand, not much success has been achieved with respect to the genes associated with clinical subpopulations, especially NPSLE. One of the reasons may be limited detection power due to small sample size. In addition, many studies thus far considered mostly the genes associated with overall SLE as candidates for the genes associated with clinical subphenotypes; however, this strategy may not necessarily be promising. Future studies should consider genome-wide case-case association analysis comparing subphenotype positive SLE and subphenotype negative SLE.

References

1. Alarcon-Segovia D, et al. Familial aggregation of systemic lupus erythematosus, rheumatoid arthritis, and other autoimmune diseases in 1,177 lupus patients from the GLADEL cohort. Arthritis Rheum. 2005;52:1138–47.
2. Deapen D, et al. A revised estimate of twin concordance in systemic lupus erythematosus. Arthritis Rheum. 1992;35:311–8.
3. Rees F, et al. The worldwide incidence and prevalence of systemic lupus erythematosus: a systematic review of epidemiological studies. Rheumatology. 2017;56:1945–61.
4. Langefeld CD, et al. Transancestral mapping and genetic load in systemic lupus erythematosus. Nat Commun. 2017;8:16021.
5. Sun C, et al. High-density genotyping of immune-related loci identifies new SLE risk variants in individuals with Asian ancestry. Nat Genet. 2016;48:323–30.
6. Morris DL, et al. Genome-wide association meta-analysis in Chinese and European individuals identifies ten new loci associated with systemic lupus erythematosus. Nat Genet. 2016;48:940–6.
7. Deng Y, Updates in Lupus Genetics TBP. Curr Rheumatol Rep. 2017;19:68.
8. Raj P, et al. Regulatory polymorphisms modulate the expression of HLA class II molecules and promote autoimmunity. Elife. 2016; 5. pii: e12089.
9. Taylor KE, et al. Risk alleles for systemic lupus erythematosus in a large case-control collection and associations with clinical subphenotypes. PLoS Genet. 2011;7:e1001311.
10. Graham RR, et al. Visualizing human leukocyte antigen class II risk haplotypes in human systemic lupus erythematosus. Am J Hum Genet. 2002;71:543–53.
11. Furukawa H, et al. Human leukocyte antigens and systemic lupus erythematosus: a protective role for the HLA-DR6 alleles DRB1*13:02 and *14:03. PLoS One. 2014;9:e87792.
12. Sirikong M, et al. Association of HLA-DRB1*1502-DQB1*0501 haplotype with susceptibility to systemic lupus erythematosus in Thais. Tissue Antigens. 2002;59:113–7.
13. Lu LY, et al. Molecular analysis of major histocompatibility complex allelic associations with systemic lupus erythematosus in Taiwan. Arthritis Rheum. 1997;40:1138–45.
14. Oka S, et al. Protective effect of the HLA-DRB1*13:02 allele in Japanese rheumatoid arthritis patients. PLoS One. 2014;9:e99453.

15. Kawasaki A, et al. Protective role of HLA-DRB1*13:02 against microscopic Polyangiitis and MPO-ANCA-positive Vasculitides in a Japanese population: a case-control study. PLoS One. 2016;11:e0154393.
16. Furukawa H, et al. Human leukocyte antigen and systemic sclerosis in Japanese: the sign of the four independent protective alleles, DRB1*13:02, DRB1*14:06, DQB1*03:01, and DPB1*02:01. PLoS One. 2016;11:e0154255.
17. Furuya T, et al. Immunogenetic features in 120 Japanese patients with idiopathic inflammatory myopathy. J Rheumatol. 2004;31:1768–74.
18. Furukawa H, et al. The role of common protective alleles HLA-DRB1*13 among systemic autoimmune diseases. Genes Immun. 2017;18:1–7.
19. Hachiya Y, et al. Association of HLA-G 3' untranslated region polymorphisms with systemic lupus erythematosus in a Japanese population: a case-control association study. PLoS One. 2016;11:e0158065.
20. Fernando MM, et al. Transancestral mapping of the MHC region in systemic lupus erythematosus identifies new independent and interacting loci at MSH5, HLA-DPB1 and HLA-G. Ann Rheum Dis. 2012;71:777–84.
21. Lintner KE, et al. Early components of the complement classical activation pathway in human systemic autoimmune diseases. Front Immunol. 2016;7:36.
22. Kim K, et al. The HLA-DRbeta1 amino acid positions 11-13-26 explain the majority of SLE-MHC associations. Nat Commun. 2014;5:5902.
23. Bennett L, et al. Interferon and granulopoiesis signatures in systemic lupus erythematosus blood. J Exp Med. 2003;197:711–23.
24. Baechler EC, et al. Interferon-inducible gene expression signature in peripheral blood cells of patients with severe lupus. Proc Natl Acad Sci U S A. 2003;100:2610–5.
25. Kawasaki A, et al. Association of IRF5 polymorphisms with systemic lupus erythematosus in a Japanese population: support for a crucial role of intron 1 polymorphisms. Arthritis Rheum. 2008;58:826–34.
26. Graham RR, et al. Three functional variants of IFN regulatory factor 5 (IRF5) define risk and protective haplotypes for human lupus. Proc Natl Acad Sci U S A. 2007;104:6758–63.
27. Kawasaki A, et al. TLR7 single-nucleotide polymorphisms in the 3' untranslated region and intron 2 independently contribute to systemic lupus erythematosus in Japanese women: a case-control association study. Arthritis Res Ther. 2011;13:R41.
28. Shen N, et al. Sex-specific association of X-linked toll-like receptor 7 (TLR7) with male systemic lupus erythematosus. Proc Natl Acad Sci U S A. 2010;107:15838–43.
29. Cunninghame Graham DS, et al. Association of NCF2, IKZF1, IRF8, IFIH1, and TYK2 with systemic lupus erythematosus. PLoS Genet. 2011;7:e1002341.
30. Bronson PG, et al. The genetics of type I interferon in systemic lupus erythematosus. Curr Opin Immunol. 2012;24:530–7.
31. Crow YJ, Manel N. Aicardi-Goutieres syndrome and the type I interferonopathies. Nat Rev Immunol. 2015;15:429–40.
32. Cuadrado E, et al. Aicardi–Goutières syndrome harbours abundant systemic and brain-reactive autoantibodies. Ann Rheum Dis. 2015;74:1931–9.
33. Abe J, et al. A nationwide survey of Aicardi-Goutieres syndrome patients identifies a strong association between dominant TREX1 mutations and chilblain lesions: Japanese cohort study. Rheumatology. 2014;53:448–58.
34. Shiozawa S, et al. Interferon-alpha in lupus psychosis. Arthritis Rheum. 1992;35:417–22.
35. Oda H, et al. Aicardi-Goutieres syndrome is caused by IFIH1 mutations. Am J Hum Genet. 2014;95:121–5.
36. Costa-Reis P, Sullivan KE. Monogenic lupus: it's all new. Curr Opin Immunol. 2017;49:87–95.
37. Ellyard JI, et al. Identification of a pathogenic variant in TREX1 in early-onset cerebral systemic lupus erythematosus by whole-exome sequencing. Arthritis Rheumatol. 2014;66:3382–6.
38. de Vries B, et al. TREX1 gene variant in neuropsychiatric systemic lupus erythematosus. Ann Rheum Dis. 2010;69:1886–7.

39. Lee-Kirsch MA, et al. Mutations in the gene encoding the 3'-5' DNA exonuclease TREX1 are associated with systemic lupus erythematosus. Nat Genet. 2007;39:1065–7.
40. Mistry P, Kaplan MJ. Cell death in the pathogenesis of systemic lupus erythematosus and lupus nephritis. Clin Immunol. 2017;185:59–73.
41. Macedo AC, Isaac L. Systemic lupus erythematosus and deficiencies of early components of the complement classical pathway. Front Immunol. 2016;7:55.
42. Sisirak V, et al. Digestion of chromatin in apoptotic cell microparticles prevents autoimmunity. Cell. 2016;166:88–101.
43. Yasutomo K, et al. Mutation of DNASE1 in people with systemic lupus erythematosus. Nat Genet. 2001;28:313–4.
44. Graham RR, et al. Genetic variants near TNFAIP3 on 6q23 are associated with systemic lupus erythematosus. Nat Genet. 2008;40:1059–61.
45. Kawasaki A, et al. Association of TNFAIP3 interacting protein 1, TNIP1 with systemic lupus erythematosus in a Japanese population: a case-control association study. Arthritis Res Ther. 2010;12:R174.
46. Sandling JK, Garnier S, Sigurdsson S, Wang C, Nordmark G, Gunnarsson I, et al. A candidate gene study of the type I interferon pathway implicates IKBKE and IL8 as risk loci for SLE. Eur J Hum Genet. 2011;1(9):479–84.
47. Lewis MJ, et al. UBE2L3 polymorphism amplifies NF-kappaB activation and promotes plasma cell development, linking linear ubiquitination to multiple autoimmune diseases. Am J Hum Genet. 2015;96:221–34.
48. Wu YY, et al. Concordance of increased B1 cell subset and lupus phenotypes in mice and humans is dependent on BLK expression levels. J Immunol. 2015;194:5692–702.
49. Samuelson EM, et al. Reduced B lymphoid kinase (Blk) expression enhances proinflammatory cytokine production and induces nephrosis in C57BL/6-lpr/lpr mice. PLoS One. 2014;9:e92054.
50. Ito I, et al. Replication of the association between the C8orf13-BLK region and systemic lupus erythematosus in a Japanese population. Arthritis Rheum. 2009;60:553–8.
51. Hom G, et al. Association of systemic lupus erythematosus with C8orf13-BLK and ITGAM-ITGAX. N Engl J Med. 2008;358:900–9.
52. Kawasaki A, et al. Role of STAT4 polymorphisms in systemic lupus erythematosus in a Japanese population: a case-control association study of the STAT1-STAT4 region. Arthritis Res Ther. 2008;10:R113.
53. Zhao J, et al. A missense variant in NCF1 is associated with susceptibility to multiple autoimmune diseases. Nat Genet. 2017;49:433–7.
54. Fredi M, et al. Typing TREX1 gene in patients with systemic lupus erythematosus. Reumatismo. 2015;67:1–7.
55. Ho RC, et al. Genetic variants that are associated with neuropsychiatric systemic lupus erythematosus. J Rheumatol. 2016;43:541–51.
56. Ota Y, et al. Single nucleotide polymorphisms of CD244 gene predispose to renal and neuropsychiatric manifestations with systemic lupus erythematosus. Mod Rheumatol. 2010;20:427–31.
57. Pullmann R Jr, et al. Apolipoprotein E polymorphism in patients with neuropsychiatric SLE. Clin Rheumatol. 2004;23:97–101.
58. Ruiz-Larranaga O, et al. Genetic association study of systemic lupus erythematosus and disease subphenotypes in European populations. Clin Rheumatol. 2016;35:1161–8.
59. Taha S, et al. Vascular endothelial growth factor G1612A (rs10434) gene polymorphism and neuropsychiatric manifestations in systemic lupus erythematosus patients. Rev Bras Reumatol Engl Ed. 2017;57:149–53.
60. Ramirez GA, et al. TRPC6 gene variants and neuropsychiatric lupus. J Neuroimmunol. 2015;288:21–4.
61. Sandrin-Garcia P, et al. Functional single-nucleotide polymorphisms in the *DEFB1* gene are associated with systemic lupus erythematosus in southern Brazilians. Lupus. 2012;21:625–31.
62. Kisiel BM, et al. Differential association of juvenile and adult systemic lupus erythematosus with genetic variants of oestrogen receptors alpha and beta. Lupus. 2011;20:85–9.

63. Yang W, et al. ITGAM is associated with disease susceptibility and renal nephritis of systemic lupus erythematosus in Hong Kong Chinese and Thai. Hum Mol Genet. 2009;18:2063–70.
64. Bassi C, et al. Efficiency of the DNA repair and polymorphisms of the XRCC1, XRCC3 and XRCC4 DNA repair genes in systemic lupus erythematosus. Lupus. 2008;17:988–95.
65. Oroszi G, et al. The Met66 allele of the functional Val66Met polymorphism in the brain-derived neurotrophic factor gene confers protection against neurocognitive dysfunction in systemic lupus erythematosus. Ann Rheum Dis. 2006;65:1330–5.
66. Liao CH, et al. Polymorphisms in the promoter region of RANTES and the regulatory region of monocyte chemoattractant protein-1 among Chinese children with systemic lupus erythematosus. J Rheumatol. 2004;31:2062–7.

Chapter 3
Immunopathology of Neuropsychiatric Systemic Lupus Erythematosus

Shunsei Hirohata

Abstract There are at least 2 separate and probably complementary pathogenetic mechanisms for NPSLE. One is the predominant ischemic-vascular involvement in large and small blood vessels, mediated mainly by anti-phospholipid antibodies (aPL), leading to focal NPSLE, such as stroke, seizures, movement disorders and myelopathy. The other is the inflammatory process with complement activation, the dysfunction of the blood-brain barrier (BBB), transudation of autoantibodies into the central nervous system (CNS) and local production proinflammatory cytokines such as IFN-α, leading to diffuse NPSLE such as psychosis, mood disorders, cognitive dysfunctions and acute confusional state. In the latter process, autoantibodies, such as anti-ribosomal P antibodies, anti-NMDA receptor NR2 antibodies and anti-Sm antibodies, play a pivotal role. Greater attention is now paid to the role of microglia in the pathogenesis.

Keywords Autoantibodies · Blood-brain barrier · Complement · Microglia · Proinflammatory cytokines

3.1 Introduction

Neuropsychiatric involvement in systemic lupus erythematosus (NPSLE) is a difficult complication of the disease, significantly impairs quality of life of the patients and results in disability or even mortality [1, 2]. Although several studies have disclosed that most frequent findings in the central nervous system (CNS) lesions are vasculopathy, microinfarction, macroinfarction, vasculitis and microthrombus, the precise mechanisms for such changes remain to be elucidated [3–5]. On the other hand, the expression of autoantibodies is a hallmark of

---

S. Hirohata (✉)
Department of Rheumatology, Nobuhara Hospital, Tatsuno, Hyogo, Japan

Department of Rheumatology & Infectious Diseases, Kitasato University School of Medicine, Sagamihara, Kanagawa, Japan

SLE. Thus, it has been recently suggested that complement deposition may play a role in the interaction between circulating autoantibodies and thromboischemic lesions observed in NPSLE [5]. On the other hand, the roles of several autoantibodies in the pathogenesis of various manifestations in NPSLE have been disclosed. In this chapter, the roles of these autoantibodies in the pathogenesis of NPSLE will be overviewed. Also special attention is directed to the mechanism of blood-brain barrier (BBB) damages, which allow various autoantibodies enter the CNS to react with neurons, in NPSLE.

3.2 Autoantibodies Implicated in the Pathogenesis of NPSLE

3.2.1 Anti-Phospholipid Antibodies

Anti-phospholipid antibodies (aPL) include anti-cardiolipin antibodies (aCL), anti-β2GP1 antibodies and lupus anticoagulant (LA). These antibodies have been found to be associated with cerebrovascular disease (thrombosis) and other manifestations including seizures, movement disorders, cognitive dysfunction and some type of myelopathy [6–9].

These antibodies have been shown to bind negatively charged phospholipids, some of which may be expressed on endothelial cell membranes [10]. Consistently, it has been revealed that IgG fractions from patients with aPL syndrome (APS) and human monoclonal aPL modulate the function of human umbilical vein endothelial cells (HUVEC) to express IL-8, MCP-1 and ICAM-1 in vitro [11]. Notably, recent studies have demonstrated that β2GPI interacts directly with TLR4 expressed on endothelial cells, contributing to β2GPI-dependent aPL-mediated endothelial cell activation [12]. Since cerebral endothelial cells regulate BBB function, it is possible that aPL might affect BBB function. In fact, it has been also reported that aPL are involved in the disruption of BBB in murine model of APS [13], although their roles in humans have not been demonstrated.

On the other hand, the direct effects on neuronal cells of aPL was suggested by in vitro studies as well as in vivo studies [14]. Thus, intracerebroventricular injection of IgG containing aPL obtained from patients induced hyperactive behavior in mice [14]. Notably, it has been demonstrated that persistent positive aPL was significantly associated with cognitive dysfunction in patients with SLE in 3 longitudinal studies [9, 15, 16]. Taking into account these findings, the mechanism for the development of cognitive dysfunction in human SLE might involve the interaction of aPL with endothelial cells as well as the direct effects of aPL on neurons. It should be pointed out, however, that intrathecally injected aPL might also react glial cells, resulting indirectly in the dysfunction of neuron.

3.2.2 Anti-Ribosomal P Antibodies

Anti-ribosomal P protein antibodies (anti-P) are directed to 3 phosphoproteins (P0, P1, and P2), which are located on the larger 60S subunit of eukaryotic ribosomes, and have molecular weights of 38, 19, and 17 kDa, respectively [17]. Anti-P have been shown to react a common linear determinant that is present in the carboxyl(C)-terminal 22-amino-acid sequence (C22) [17]. Anti-P are specific for SLE and detected in 12–16% [17]. Although the association of serum anti-P with NPSLE has been controversial in retrospective cross-sectional studies [18–22], their association with lupus psychosis (diffuse NPSLE) has been confirmed in longitudinal and prospective studies [23–27].

Although anti-P directed to the C-terminal 22-amino acids epitope have been shown to be major autoantibodies to ribosomal P proteins [17, 18, 23], the frequency of their detection in cerebrospinal fluid (CSF) from patients with diffuse NPSLE was not high enough to ensure their involvement in the pathogenesis of this disease [17, 18, 28]. Notably, CSF anti-P recognizing the ribosomal P protein epitope other than the C-terminal 22-amino acids, have been shown to be elevated in patients with diffuse NPSLE [29].

The presence of an epitope that is antigenically related to the C-terminal amino acids of ribosomal P proteins has been demonstrated on the surfaces of a variety of cells, including human hepatoma cells, neuroblastoma cells, fibroblasts, and endothelial cells [30]. In healthy mice, anti-P were found to react with neurons in the hippocampus, cingulate and primary olfactory piriform cortex [14]. Accordingly, intracerebroventricular injection of anti-P induced a long-term depression and smell deficits in healthy mice [14]. On the other hand, it has been demonstrated that antibodies reactive with the C-terminal 11 amino acids of ribosomal P proteins [31] from SLE patients induced calcium influx and apoptosis in rat neurons, but not in astrocytes, in vitro [32]. Furthermore, injection of these antibodies into the brain of living rats also triggered neuronal death by apoptosis [32]. It was found that these antibodies reactive with the C-terminal 11 amino acids of ribosomal P proteins were targeted against p331 present in rat brain synaptosomal fractions highly enriched in synaptic regions [33]. Thus, p331 was termed neuronal surface P antigens (NSPAs), which is preferentially distributed in areas involved in memory, cognition, and emotion [32].

Previous studies have disclosed that the ribosomal P epitope is also expressed on the surfaces of activated human peripheral blood CD4+ T cells [34] and monocytes [35], but not on the surface of resting or activated B cells [34]. More interestingly, it has been shown that anti-P enhanced the expression of protein and mRNA for TNF-α and IL-6 in activated monocytes [35]. Thus, it is suggested that anti-P might modify a variety of inflammatory responses through upregulation of the expression of proinflammatory cytokines in monocytes, possibly resulting in the disruption of BBB.

3.2.3 Anti-NMDA Receptor NR2 Subunit Antibodies

N-methyl-D-aspartate (NMDA) receptors are a subgroup of the glutamate receptor family, responsible for the majority of excitatory synaptic transmission in the CNS [36]. It has been shown that injection of anti-NMDA receptor NR2 antibodies (anti-NR2) purified from the sera or CSF of patients with diffuse NPSLE into mice brain resulted in apoptosis of the neuronal cells [37]. More importantly, mice induced by antigen to express anti-NR2 in sera had no neuronal damage until breakdown of the BBB was induced, confirming that direct contact of anti-NR2 with neurons is indispensable for the neuronal damages [38].

Consistently, in humans, CSF anti-NR2 was elevated in diffuse NPSLE compared with that in focal NPSLE or in non-SLE control, whereas there was no significant difference in serum anti-NR2 among the 3 groups [39]. Several studies have reported that anti-NR2 are involved in cognitive dysfunction and mood disorders in SLE [40, 41]. Furthermore, a synergism between anti-NR2 and aPL has been implicated in inducing tissue damage and cognitive dysfunction [42].

Of interest, the effect of anti-NR2 on neurons was shown to be dose dependent [43]. Thus, at low concentrations they alter synaptic function, whereas at higher concentrations they can cause neuronal cell death by apoptosis [43]. Notably, CSF anti-NR2 levels were the highest in acute confusional state (ACS), the severest form of diffuse NPSLE, among various types of NPSLE [44]. It is therefore suggested that higher concentrations of anti-NR2 within the CNS might result in stronger neuronal damages, leading to the development of ACS. In ACS, the breach of BBB is considered to play a critical role in the elevation of CSF anti-NR2 levels [44]. In fact, CSF anti-NR2 levels were significantly correlated with Q albumin in diffuse NPSLE [44]. It should be pointed out that anti-NR2 react with endothelial cells and upregulate the production of IL-6 and IL-8 through activation of NFkB [45]. It is therefore possible that anti-NR2 might be involved in the disruption of BBB as well as in the development of vasculitis in SLE [45]. In this regard, the biological effects of anti-NR2 are comparable with those of anti-P.

The presence of autoantibodies to the conformational structure of NR1/NR2 heterodimer has been discovered in sera from patients with ovarian teratoma [46]. Such antibodies result in the development of various neurological manifestations, called NMDA encephalitis [47]. We have found that the expression of these antibodies was rare even in NPSLE patients with high titers of anti-NR2 [unpublished observation]. Thus, it is conceivable that the mechanism of expression of anti-NMDA receptor NR1/NR2 might be different from that of anti-NMDA receptor NR2.

3.2.4 Anti-Sm Antibodies

The association of serum anti-Sm antibodies with CNS involvement in SLE was previously suggested [48, 49]. Notably, a strong association was found between serum anti-Sm and organic brain syndrome, consisting mainly of ACS of diffuse

NPSLE [50]. However, the precise mechanism by which anti-Sm causes diffuse NPSLE has remained unclear until recently.

We have recently disclosed that CSF anti-Sm is elevated in ACS of diffuse NPSLE, whereas there were no significant differences in CSF anti-RNP among various subtypes of NPSLE, including ACS, non-ACS diffuse NPSLE and focal NPSLE [51]. Of importance is the observation that monoclonal anti-Sm as well as purified human anti-Sm bound to the surface of SK-N-MC cells, confirming that the epitopes recognized by anti-Sm exist on the surface of neuronal cells [51]. It is therefore suggested that the presence of higher concentrations of anti-Sm within the CNS might cause more extensive neuronal damages, leading to the development of ACS, as is in the case with anti-NR2 [44].

As is in the case with CSF anti-NR2 in ACS [44], the elevations of CSF anti-Sm levels in ACS compared with those in non-ACS diffuse NPSLE or in focal NPSLE could not be accounted for by the increased intrathecal synthesis of anti-Sm, but most likely resulted from the breach of BBB, as evidenced by the elevation of Q albumin [51]. In fact, CSF anti-Sm were significantly correlated with Q albumin (marker for damages of BBB) and with CSF anti-NR2 in patients with NPSLE [51]. The data therefore indicate that the elevation of both anti-Sm and anti-NR2 in CSF plays a crucial role in the development of ACS [51]. Further studies would be interesting to explore whether there might be any synergistic effects between anti-Sm and anti-NR2.

It should be pointed out that the association of serum anti-Sm with ACS has been confirmed in a different set of patients with NPSLE [51] independently of the previous study [50]. It is therefore likely that serum anti-Sm might play an additional role in the pathogenesis of ACS other than providing anti-Sm into the CNS to contact directly with neuronal cells. Thus, anti-Sm might influence the functions of cells present not only in the CNS, but in the systemic circulation.

3.2.5 Anti-Neuronal Antibodies

The role of anti-neuronal antibodies in the pathogenesis of NPSLE has been well appreciated since it was demonstrated that IgG anti-neuronal antibodies were present at much higher concentrations in CSF than in sera from patients with active NPSLE by Bluestein et al. [52]. Notably, CSF IgG antineuronal antibodies were significantly elevated in patients with diffuse NPSLE compared with focal NPSLE, such as neurologic syndromes [28]. As to the epitopes to which CSF anti-neuronal antibodies were directed, all of aPL, anti-P, anti-NR2 and anti-Sm have been found to bind neuronal cells. In addition, autoantibodies to NMDA receptor NR1 subunit have been shown to be also increased in CSF from patients with diffuse NPSLE [53]. Therefore, anti-neuronal antibodies might bind so many targets that the nature of anti-neuronal antibodies might be different from disease to disease, or from patient to patient. It would be therefore necessary to study antibodies whose target molecules are identified.

Iizuka et al. tried to identify the target molecules to which autoantibodies reactive with SK-N-MC cells are directed by proteomics analysis [54]. They identified 4 proteins, including peroxiredoxin-4, ubiquitin carboxyl-terminal hydrolase isozyme L1, splicing factor arginine/serine-rich 3, and histone H2A type 1 as candidate autoantigens for the anti-neuronal antibodies [54]. However, their roles in the pathogenesis of NPSLE remain to be explored.

3.2.6 Other Autoantibodies

Anti-glycolipid antibodies, including anti-asialo GM1, were demonstrated to be present in the sera of SLE patients [55, 56]. However, these antibodies were also found in patients with cerebral trauma, and other non-autoimmune neurological disorders [24]. It is therefore most likely that these antibodies might be produced as a result of the leakage of brain antigens into the systemic circulation after brain injury.

Microtubule-associated protein 2 (MAP-2) is a cellular protein required for the control of cytoskeletal integrity and other neuronal functions, which is found almost exclusively in neurons [57]. Anti-MAP-2 antibodies were found in 17% of patients with SLE in relation with neuropsychiatric symptoms, such as psychosis, seizure, neuropathy and cerebritis [57]. Therefore, anti-MAP 2 antibodies are one of the constituents of anti-neuronal antibodies.

Anti-aquaporin 4 (AQP4) IgG autoantibodies, which have been found to be associated with neuromyelitis optica spectrum disorders (NMOSD), were detected in a fraction of patients with NPSLE, especially demyelinating disorder [58]. It should be noted, however, that AQP4 is present on astrocytes, but not on neuron. It is also evident that the breach of BBB is required for anti-AQP4 antibodies to reach astrocytes.

3.3 Intrathecal Ig Production and Blood-Brain Barrier Damages in NPSLE

3.3.1 Intrathecal Ig Production

In NPSLE, autoantibodies levels in CSF have been shown to be more closely correlated with neuropsychiatric manifestations than those in serum [39, 44, 59]. Two mechanisms have been implicated for the elevation of CSF IgG, including intrathecal IgG production and transudation through the disrupted BBB [60]. Previous studies disclosed that CSF IgG index, an indicator of intrathecal IgG synthesis, is elevated in NPSLE [61, 62]. On the other hand, anti-NR2 and anti-Sm in CSF have been shown to result in ACS [44, 51]. Although the intrathecal production of these autoantibodies (CSF anti-NR2 index and CSF anti-Sm index) was elevated in NPSLE compared with non-SLE control, there were no significant differences among the 3 groups of NPSLE, including ACS, non-ACS diffuse NPSLE and focal NPSLE [44, 51]. It should be pointed out that the elevation of CSF IgG index was not confined to diffuse NPSLE [61, 62]. Thus, it makes sense that CSF IgG index is elevated in focal NPSLE as well as in ACS and non-ACS NPSLE. Elevated CSF IgG index is usually associated with CSF oligoclonal IgG bands [61]. However, the presence of CSF oligoclonal IgG bands was not confined to diffuse NPSLE [61]. Thus, it is strongly suggested that the enhanced intrathecal IgG production might not be specific for diffuse NPSLE, but a common feature in NPSLE, including focal NPSLE.

3.3.2 Blood-Brain Barrier Damages in NPSLE

There has been accumulating evidence for the role of BBB dysfunction in the pathogenesis of NPSLE [64]. It has been well appreciated that Q albumin ([CSF albumin / serum albumin] × 1000) is a reliable marker for the integrity of BBB, since albumin is produced exclusively in the liver, but not in the CNS [60]. Recent studies have demonstrated that Q albumin was significantly higher in ACS than in non-ACS diffuse NPSLE, indicating that the breach of BBB plays a crucial role in the development of ACS allowing the increased migration of anti-NR2 and anti-Sm into the CNS [44, 51]. Moreover, CSF anti-Sm was significantly correlated with CSF anti-NR2 in NPSLE [44, 51]. Since there was no significant correlation between serum anti-Sm and anti-NR2, the positive correlation between CSF anti-Sm and anti-NR2 might be accounted for by the BBB disruption [44, 51].

The mechanism of damages in BBB in NPSLE has not been well documented at present. It is likely that several autoantibodies, including as anti-P and anti-NR2, might result in BBB damage, since they react with endothelial cells [30, 45]. In fact, anti-NR2 bound to endothelial cells to produce IL-6 and IL-8 through NFkB activation [45]. In addition, anti-P have been shown to upredulate the expression of proinflammatory cytokines in monocytes [35], which may affect the function of BBB. Of interest, recent studies have disclosed that anti-Sm synergized with hemoglobin to enhance the secretion of proinflammatory cytokines while eliciting the increased production of monocyte migratory signals from endothelial cells [63]. It is therefore also possible that anti-Sm also might result in BBB damage.

It is suggested that endothelial cells might be stimulated by proinflammatory cytokines or autoantibodies to up-regulate the expression of adhesion molecules on their surface promoting entry of lymphocytes into the CNS [65]. In fact, serum levels of soluble ICAM-1 increase along with systemic disease activity in patients with SLE, and normalize with remission [66]. It is therefore likely that activated endothelial cells might lead to an impaired integrity of the BBB as a prerequisite for CNS disease activity [67].

A complement split product C5a has been shown to alter BBB integrity in MRL/lpr mice through activation of NFkB [68, 69]. Notably, recent studies have demonstrated that C5a decreases the expression of tight junction proteins, claudin 5, ZO-1, and occludin, in human brain microvascular endothelial cells, possibly leading to the breach of BBB [70]. On the other hand, it has been disclosed that serum C5a levels are elevated in NPSLE, possibly through acceleration of the conversion from C5 [71]. Thus, the elevation of serum C5a might account for the impaired BBB function in NPSLE.

3.4 Roles of Complements and Microglia in the Pathogenesis of NPSLE

3.4.1 Complements

A link between complement-triggered pathways and the regulation of developmental synaptic pruning has been reported. Thus, Stevens B et al. have identified that C1q and C3 are key contributors to synaptic refinement and neuronal connectivity in the CNS [72]. Recent studies have disclosed that BBB disruption resulted in the elevated levels of IL-6 and C3 in CSF in diffuse NPSLE, especially in ACS [73]. It is therefore suggested that the elevation of C3 in CSF might affect the synaptic functions of neurons. Of interest, mechanistic evidence of pathogenic crosstalk between complement and reactive microglia in the CNS has been recently shown. Thus, it is revealed that complement plays a pivotal role in the induction of neurotoxic astrocytes in various neurodegenerative diseases [74].

The expression of C5a receptor (C5aR) has been recently demonstrated on human hepatocytes, epithelial cells, endothelial cells, and tissue mast cells [75, 76]. Furthermore, although the expression of C5aR in the normal brain is very low, C5aR expression is greatly upregulated on reactive astrocytes, microglia, neurons, and endothelial cells in inflamed brain [75]. In this regard, it is likely that C5a might also affect the functions of glial cells and neurons during the active inflammation [77].

3.4.2 Microglia

Microglia, resident macrophages of the CNS, display increased phagocytic activity and inflammatory cytokine production in CNS inflammation or damage [78]. Notably, such reactive microglia have been detected in some lupus model mouse [79]. Of note, microglia have been found to be sensitive to type I interferon (IFN) to be activated [80]. SLE is characterized by the expression of type I IFN-induced genes [81]. In fact, type I IFN was found to be elevated in CSF from patients with NPSLE [82]. Recent studies have demonstrated that type I IFN stimulates microglia to be activated to engulf neuronal and synaptic material, resulting synaptic loss, in lupus-prone mice [83]. Furthermore, it has been demonstrated that IFN-α receptor 1 (IFNAR1) signaling was elevated, as evidenced by the expression of MXA (encoded by MX1), in microglia of patients with NPSLE [83].

On the other hand, it has been revealed that exposure to SLE patient sera resulted in morphological changes in the microglia along with an increase in the expression of MHC II antigen and CD86 and the release of nitric oxide and proinflammatory cytokines [84]. Since inactivating complements or neutralizing proinflammatory cytokines of the sera did not abrogate their ability of microglial activation, it is suggested that some antibodies might be involved in the activation of microglia [84]. Whether type I IFN was involved in the activation of microglia by SLE sera needs to be explored.

3.5 Summary

In general, autoantibodies play pivotal roles in the pathogenesis of NPSLE, as highlighted in Table 3.1 in which association of each autoantibody with specific neuropsychiatric manifestations is indicated. Thus, aPL are important in the development of focal NPSLE, including cerebrovascular diseases, seizures, movement disorders and myelopathy. On the other hand, several autoantibodies, such as anti- P, anti- NR2 and anti-Sm, play a pivotal role in the pathogenesis of diffuse NPSLE, including psychosis, mood disorders, cognitive dysfunctions and

Table 3.1 Autoantibodies involved in the development of various neuropsychiatric manifestations in NPSLE

Neuropsychiatric manifestations	Serum antibodies	CSF antibodies
Diffuse NPSLE		
Acute confusional state	Anti-Sm, Anti-P	Anti-Sm, Anti-NR2, Anti-P
Anxiety disorders	Anti-P	Anti-NR2, Anti-P
Cognitive dysfunctions	Anti-P, Anti-PL	Anti-NR2, Anti-P
Mood disorders	Anti-P	Anti-NR2, Anti-P
Psychosis	Anti-P	Anti-NR2, Anti-P
Focal NPSLE		
Aseptic meningitis	Anti-RNP	
Cerebrovascular diseases	Anti-PL	
Demyelinating syndrome	Anti-AQP4	
Headache	Anti-PL	
Movement disorder	Anti-PL	
Myelopathy	Anti-PL, Anti-NR2, Anti-AQP4	
Seizure disorder	Anti-PL	

Anti-P anti-ribosomal P, Anti-PL anti-phospholipid, Anti-NR2 anti-NMDA receptor NR2, Anti-AQP4 anti-aquaporin 4

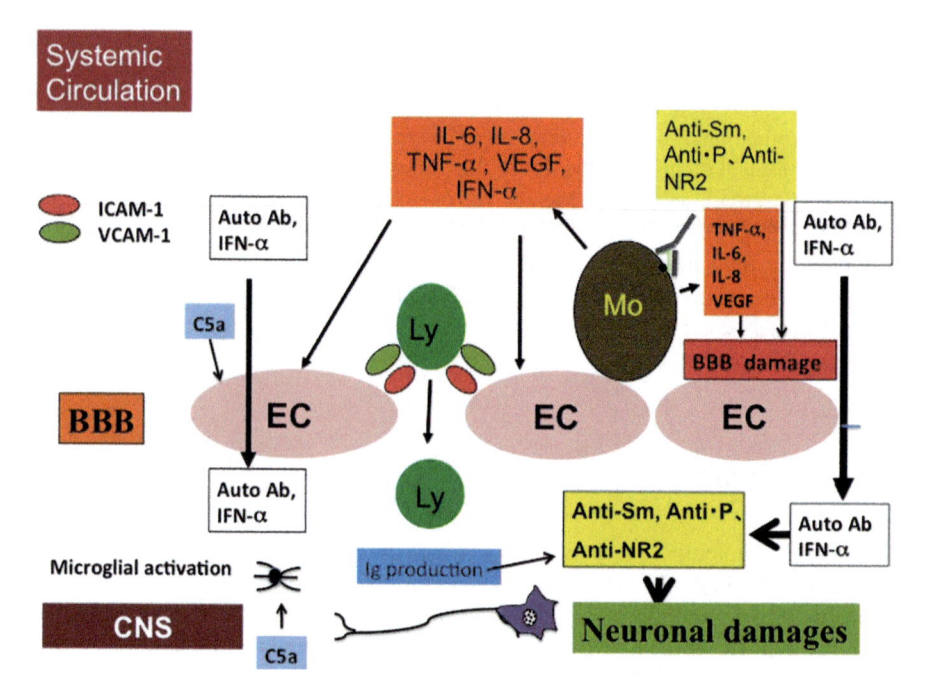

Fig. 3.1 Suggested pathogenesis of diffuse NPSLE. Various autoantibodies and cytokines are involved. C5a also plays a role not only in the breach of blood-brain barrier (BBB) but in the damages of neuron through microglia. IFN-α also causes damages of neuron through microglia

ACS. The disruption of BBB function is critical for the development of diffuse NPSLE, especially ACS, in that it allows neuron-reactive autoantibodies to enter the CNS to react with neurons. Recently, the roles of complement not only in the disruption of BBB but in the dysfunction of neurons have been implicated. Furthermore, greater attention is now paid to the roles of microglia, in relation with type I IFN, in the pathogenesis of NPSLE. Taking these findings into account, the suggested pathogenesis of diffuse NPSLE is depicted in Fig. 3.1.

References

1. Gibson T, Myers AR. Nervous system involvement in systemic lupus erythematosus. Ann Rheum Dis. 1975;35:398–406.
2. Harris EN, Hughes GR. Cerebral disease in systemic lupus erythematosus. Springer Semin Immunopathol. 1985;8:251–66.
3. Ellis SG, Verity MA. Central nervous system involvement in systemic lupus erythematosus: a review of neuropathologic findings in 57 cases, 1955–1977. Semin Arthritis Rheum. 1979;8:212–21.
4. Hanly JG, et al. Brain pathology in systemic lupus erythematosus. J Rheumatol. 1992;19:732–41.

5. Cohen D, et al. Brain histopathology in patients with systemic lupus erythematosus: identification of lesions associated with clinical neuropsychiatric lupus syndromes and the role of complement. Rheumatology. 2017;55:77–86.
6. Sanna G, et al. Neuropsychiatric manifestations in systemic lupus erythematosus: prevalence and association with antiphospholipid antibodies. J Rheumatol. 2003;30:985–92.
7. Andrade RM, et al., LUMINA Study Group. Seizures in patients with systemic lupus erythematosus: data from LUMINA, a multiethnic cohort (LUMINA LIV). Ann Rheum Dis. 2008; 67:829–34.
8. Appenzeller S, et al. Epileptic seizures in systemic lupus erythematosus. Neurology. 2004;63:1808–12.
9. McLaurin EY, et al. Predictors of cognitive dysfunction in patients with systemic lupus erythematosus. Neurology. 2005;64:297–303.
10. Harris EN, et al. Cross-reactivity of antiphospholipid antibodies. J Clin Lab Immnnol. 1985;16:1–6.
11. Clemens N, et al. In vitro effects of antiphospholipid syndrome-IgG fractions and human monoclonal antiphospholipid IgG antibody on human umbilical vein endothelial cells and monocytes. Ann N Y Acad Sci. 2009;1173:805–13.
12. Raschi E, et al. β2-glycoprotein I, lipopolysaccharide and endothelial TLR4: three players in the two hit theory for anti-phospholipid-mediated thrombosis. J Autoimmun. 2014;55:42–50.
13. Katzav A, et al. The pathogenesis of neural injury in animal models of the antiphospholipid syndrome. Clin Rev Allergy Immunol. 2010;38:196–200.
14. Katzav A, et al. Antibody-specific behavioral effects: intracerebroventricular injection of antiphospholipid antibodies induces hyperactive behavior while anti-ribosomal-P antibodies induces depression and smell deficits in mice. J Neuroimmunol. 2014;272:10–5.
15. Hanly JG, et al. A prospective analysis of cognitive function and anticardiolipin antibodies in systemic lupus erythematosus. Arthritis Rheum. 1999;42:728–34.
16. Menon S, et al. A longitudinal study of anticardiolipin antibody levels and cognitive functioning in systemic lupus erythematosus. Arthritis Rheum. 1999;42:735–41.
17. Elkon K, et al. Identification and chemical synthesis of a ribosomal protein antigenic determinant in systemic lupus erythematosus. Proc Natl Acad Sci U S A. 1986;83:7419–23.
18. Schneebaum AB, et al. Association of psychiatric manifestations with antibodies to ribosomal P proteins in systemic lupus erythematosus. Am J Med. 1991;90:54–6.
19. Nojima Y, et al. Correlation of antibodies to ribosomal P protein with psychosis in patients with systemic lupus erythematosus. Ann Rheum Dis. 1992;51:1053–5.
20. Isshi K, Hirohata S. Association of anti-ribosomal P protein antibodies with neuropsychiatric systemic lupus erythematosus. Arthritis Rheum. 1996;39:1483–90.
21. Karassa FB, et al. Accuracy of anti-ribosomal P protein antibody testing for the diagnosis of neuropsychiatric systemic lupus erythematosus: an international meta-analysis. Arthritis Rheum. 2006;4:312–24.
22. Haddouk S, et al. Clinical and diagnostic value of ribosomal P autoantibodies in systemic lupus erythematosus. Rheumatology. 2009;48:953–7.
23. Bonfa E, et al. Association between lupus psychosis and antiribosomal P protein antibodies. N Engl J Med. 1987;317:265–71.
24. West SG, et al. Neuropsychiatric lupus erythematosus: a 10-year prospective study on the value of diagnostic tests. Am J Med. 1995;99:153–63.
25. Watanabe T, et al. Neuropsychiatric manifestations in patients with systemic lupus erythematosus: diagnostic and predictive value of longitudinal examination of anti-ribosomal P antibody. Lupus. 1996;5:178–83.
26. Briani C, et al. Neurolupus is associated with anti-ribosomal P protein antibodies: an inception cohort study. J Autoimmun. 2009;32:79–84.
27. Hanly JG, et al., Systemic Lupus International Collaborating Clinics. Autoantibodies and neuropsychiatric events at the time of systemic lupus erythematosus diagnosis: results from an international inception cohort study. Arthritis Rheum. 2008; 58:843–53.

28. Isshi K, Hirohata S. Differential roles of the anti-ribosomal P antibody and antineuronal antibody in the pathogenesis of central nervous system involvement in systemic lupus erythematosus. Arthritis Rheum. 1998;41:1819–27.
29. Hirohata S, et al. Association of cerebrospinal fluid anti-ribosomal P protein antibodies with diffuse psychiatric/neuropsychological syndromes in systemic lupus erythematosus. Arthritis Res Ther. 2007;9:R44.
30. Koren E, et al. Autoantibodies to the ribosomal P proteins react with a plasma membrane-related target on human cells. J Clin Invest. 1992;89:1236–41.
31. Elkon K, et al. Properties of the ribosomal P2 protein autoantigen are similar to those of foreign protein antigens. Proc Natl Acad Sci U S A. 1988;85:5186–9.
32. Matus S, et al. Antiribosomal-P autoantibodies from psychiatric lupus target a novel neuronal surface protein causing calcium influx and apoptosis. J Exp Med. 2007;204:3221–34.
33. Nagy A, Delgado-Escueta AV. Rapid preparation of synaptosomes from mammalian brain using nontoxic isoosmotic gradient material (Percoll). J Neurochem. 1984;43:1114–23.
34. Hirohata S, Nakanishi K. Antiribosomal P protein antibody in human systemic lupus erythematosus reacts specifically with activated T cells. Lupus. 2001;10:612–21.
35. Nagai T, et al. Anti-ribosomal P protein antibody in human systemic lupus erythematosus up-regulates the expression of proinflammatory cytokines by human peripheral blood monocytes. Arthritis Rheum. 2005;52:847–55.
36. Furukawa H, et al. Subunit arrangement and function in NMDA receptors. Nature. 2005;438:185–92.
37. DeGiorgio LA, et al. A subset of lupus anti-DNA antibodies cross-reacts with the NR2 glutamate receptor in systemic lupus erythematosus. Nat Med. 2001;7:1189–93.
38. Kowal C, et al. Cognition and immunity; antibody impairs memory. Immunity. 2004;21:179–88.
39. Arinuma Y, et al. Association of cerebrospinal fluid anti-NR2 glutamate receptor antibodies with diffuse neuropsychiatric systemic lupus erythematosus. Arthritis Rheum. 2008;58:1130–5.
40. Harrison MJ, et al. Relationship between serum NR2a antibodies and cognitive dysfunction in systemic lupus erythematosus. Arthritis Rheum. 2006;54:2515–22.
41. Lapteva L, et al. Anti-N-methyl-D-aspartate receptor antibodies, cognitive dysfunction, and depression in systemic lupus erythematosus. Arthritis Rheum. 2006;54:2505–14.
42. Gerosa M, et al. Antiglutamate receptor antibodies and cognitive impairment in primary antiphospholipid syndrome and systemic lupus erythematosus. Front Immunol. 2016;7:5.
43. Faust TW, et al. Neurotoxic lupus autoantibodies alter brain function through two distinct mechanisms. Proc Natl Acad Sci U S A. 2010;107:18569–74.
44. Hirohata S, et al. Blood-brain barrier damages and intrathecal synthesis of anti-N-methyl-D-aspartate receptor NR2 antibodies in diffuse psychiatric/neuropsychological syndromes in systemic lupus erythematosus. Arthritis Res Ther. 2014;16:R77.
45. Yoshio T, et al. IgG anti-NR2 glutamate receptor autoantibodies from patients with systemic lupus erythematosus activate endothelial cells. Arthritis Rheum. 2013;65:457–63.
46. Iizuka T, et al. Anti-NMDA receptor encephalitis in Japan: long-term outcome without tumor removal. Neurology. 2008;70:504–11.
47. Graus F, et al. A clinical approach to diagnosis of autoimmune encephalitis. Lancet Neurol. 2016;15:391–404.
48. Winfield JB, et al. Serologic studies in patients with systemic lupus erythematosus and central nervous system dysfunction. Arthritis Rheum. 1978;21:289–94.
49. Yasuma M, et al. Clinical significance of IgG anti-Sm antibodies in patients with systemic lupus erythematosus. J Rheumatol. 1990;17:469–75.
50. Hirohata S, Kosaka M. Association of anti-Sm antibodies with organic brain syndrome secondary to systemic lupus erythematosus. Lancet. 1994;343:796.
51. Hirohata S, et al. Association of cerebrospinal fluid anti-Sm antibodies with acute confusional state in systemic lupus erythematosus. Arthritis Res Ther. 2014;16:450.
52. Bluestein HG, et al. Cerebrospinal fluid antibodies to neuronal cells: association with neuropsychiatric manifestations of systemic lupus erythematosus. Am J Med. 1981;70:240–6.

53. Ogawa E, et al. Association of antibodies to the NR1 subunit of N-methyl-D-aspartate receptors with neuropsychiatric systemic lupus erythematosus. Mod Rheumatol. 2016;26:377–83.
54. Iizuka N, et al. Identification of autoantigens specific for systemic lupus erythematosus with central nervous system involvement. Lupus. 2010;19:717–26.
55. Hirano T, et al. Antiglycolipid autoantibody detected in the sera from systemic lupus erythematosus patients. J Clin Invest. 1980;66:1437–40.
56. Endo T, et al. Antibodies to glycosphingolipids in patients with multiple sclerosis and SLE. J Immunol. 1984;132:1793–7.
57. Williams RC, et al. Antibodies to microtubule-associated protein 2 in patients with neuropsychiatric systemic lupus erythematosus. Arthritis Rheum. 2004;50:1239–47.
58. Mader S, et al. Understanding the antibody repertoire in neuropsychiatric systemic lupus erythematosus and neuromyelitis optica spectrum disorders: do they share common targets? Arthritis Rheumatol. 2018;70:277–86.
59. Yoshio T, et al. Antiribosomal P protein antibodies in cerebrospinal fluid are associated with neuropsychiatric systemic lupus erythematosus. J Rheumatol. 2005;32:34–9.
60. Tibbling G, et al. Principles of albumin and IgG analyses in neurological disorders. I. Establishment of reference values. Scand J Clin Lab Invest. 1977;37:385–90.
61. Winfield JB, et al. Intrathecal IgG synthesis and blood-brain barrier impairment in patients with systemic lupus erythematosus and central nervous system dysfunction. Am J Med. 1983;74:837–44.
62. Hirohata S, et al. Cerebrospinal fluid IgM, IgA, and IgG indexes in systemic lupus erythematosus. Their use as estimates of central nervous system disease activity. Arch Intern Med. 1985;145:1843–6.
63. Bhatnagar H, et al. Serum and organ- associated anti-hemoglobin humoral autoreactivity: association with anti-Sm responses and inflammation. Eur J Immunol. 2011;41:537–48.
64. Abbott NJ, et al. The blood-brain barrier in systemic lupus erythematosus. Lupus. 2003;12:908–15.
65. Zaccagni H, et al. Soluble adhesion molecule levels, neuropsychiatric lupus and lupus-related damage. Front Biosci. 2004;9:1654–9.
66. Spronk PE, et al. Levels of soluble VCAM-1, soluble ICAM-1, and soluble E-selectin during disease exacerbations in patients with systemic lupus erythematosus (SLE); a long term prospective study. Clin Exp Immunol. 1994;97:439–44.
67. Ainiala H, et al. Increased serum metalloproteinase 9 levels in systemic lupus erythematosus patients with neuropsychiatric manifestations and brain magnetic resonance imaging abnormalities. Arthritis Rheum. 2004;50:858–65.
68. Jacob A, et al. C5a alters blood-brain barrier integrity in experimental lupus. FASEB J. 2010;24:1682–8.
69. Jacob A, et al. C5a/CD88 signaling alters blood-brain barrier integrity in lupus through nuclear factor-kappa B. J Neurochem. 2011;119:1041–51.
70. Mahajan SD, et al. C5a alters blood-brain barrier integrity in a human in vitro model of systemic lupus erythematosus. Immunology. 2015;146:130–43.
71. Sakuma Y, et al. Differential activation mechanisms of serum C5a in lupus nephritis and neuropsychiatric systemic lupus erythematosus. Mod Rheumatol. 2017;27:292–7.
72. Stevens B, et al. The classical complement cascade mediates CNS synapse elimination. Cell. 2007;131:1164–78.
73. Asano T, et al. Evaluation of blood-brain barrier function by quotient alpha2 macroglobulin and its relationship with interleukin-6 and complement component 3 levels in neuropsychiatric systemic lupus erythematosus. PLoS One. 2017;12:e0186414.
74. Liddelow SA, et al. Neurotoxic reactive astrocytes are induced by activated microglia. Nature. 2017;541:481–7.
75. Gasque P, et al. Expression of the receptor for complement C5a (CD88) is up-regulated on reactive astrocytes, microglia, and endothelial cells in the inflamed human central nervous system. Am J Pathol. 1997;150:31–41.

76. Laudes IJ, et al. Expression and function of C5a receptor in mouse microvascular endothelial cells. J Immunol. 2002;169:5962–70.
77. Woodruff TM, et al. The role of the complement system and the activation fragment C5a in the central nervous system. NeuroMolecular Med. 2010;12:179–92.
78. Lynch MA. The multifaceted profile of activated microglia. Mol Neurobiol. 2009;40:139–56.
79. Mondal TK, et al. Autoantibody-mediated neuroinflammation: pathogenesis of neuropsychiatric systemic lupus erythematosus in the NZM88 murine model. Brain Behav Immun. 2008;22:949–59.
80. Goldmann T, et al. USP18 lack in microglia causes destructive interferonopathy of the mouse brain. EMBO J. 2015;34:1612–29.
81. Kirou KA, et al. Coordinate overexpression of interferon-α -induced genes in systemic lupus erythematosus. Arthritis Rheum. 2004;50:3958–67.
82. Shiozawa S, et al. Interferon-alpha in lupus psychosis. Arthritis Rheum. 1992;35:417–22.
83. Bialas AR, Presumey J, Das A, et al. Microglia-dependent synapse loss in type I interferon-mediated lupus. Nature. 2017;546:539–43.
84. Wang J, et al. Microglia activation induced by serum of SLE patients. J Neuroimmunol. 2017;310:135–42.

Chapter 4
Pathology of Neuropsychiatric Systemic Lupus Erythematosus

Shunsei Hirohata

Abstract Small vessel non-inflammatory vasculopathy, microvessel occlusion, multifocal microinfarcts, microhaemorrhages and cortical atrophy were most frequently observed pathological changes in NPSLE. Although vasculitis in the brain is rather uncommon, it might be detected in about 10% of patients with NPSLE. It has been revealed that activation of complement appears to play an important role in the development of microvasculopathy in NPSLE. Autoantibodies, including anti-phospholipid antibodies, causing direct injury of endothelial cells would be involved in the deposition of complements. Since anti-NMDA receptor NR2 antibodies have been demonstrated to react with endothelial cells to induce the production of inflammatory cytokines, it is possible that these antibodies might also result in the microvascular changes in the brain. Recent studies have shed light on the roles of microglia in relation with type I interferons in the pathogenesis of NPSLE.

Keywords Pathology · Microvasculopathy · Vasculitis · Autoantibodies · Magnetic resonance imaging

4.1 Introduction

Neuropsychiatric involvement in systemic lupus erythematosus (NPSLE) is one of the recalcitrant complications of the disease, leading to substantial impairment of quality of life as well as disability [1, 2]. A variety of neuropsychiatric manifestations are seen in patients with SLE. In general, there are 2 main pathogenic mechanisms for NPSLE: one is an ischemic-vascular change in large and small blood vessels, mediated mainly by anti-phospholipid antibodies (aPL), leading to focal manifestations (focal NPSLE), whereas the other is an inflammatory-neurotoxic

S. Hirohata (✉)
Department of Rheumatology, Nobuhara Hospital, Tatsuno, Hyogo, Japan

Department of Rheumatology & Infectious Diseases, Kitasato University School of Medicine, Sagamihara, Kanagawa, Japan

© Springer International Publishing AG, part of Springer Nature 2018
S. Hirohata (ed.), *Neuropsychiatric Systemic Lupus Erythematosus*,
https://doi.org/10.1007/978-3-319-76496-2_4

change mediated by autoantibodies entering into the central nervous system (CNS) through the breach of the blood-brain barrier (BBB), usually occurring as psychiatric manifestations (diffuse NPSLE) [3].

In this chapter, the overall pathological features in NPSLE will be overviewed.

4.2 Overall Characteristic Features in Pathology in NPSLE

Macroscopic pathologic features are generally reflected on MRI scans. Abnormalities on MRI scans are observed in 47% of patients with NPSLE [4]. The majority of such abnormalities was white matter high intensity lesions (WMHIs), grey matter high intensity lesions (GMHIs) and brain atrophy [4]. Accordingly, at autopsy widespread abnormalities of cerebral vasculature have been found in patients with NPSLE [5, 6]. Thus, small vessel non-inflammatory vasculopathy, microvessel occlusion, multifocal microinfarcts, microhaemorrhages and cortical atrophy were most commonly observed features [7–10].

Neurological manifestations such as stroke, cortical visual defects, cranial nerve abnormalities, and in some cases, seizures may be caused by structural lesions due to vasculopathy. However, these findings were not confined to NPSLE [10]. By contrast, a few patients with NPSLE have almost no vascular abnormalities at autopsy [5, 6]. Of note, true vasculitis characterized by cellular infiltration in the walls is rare in NPSLE even in the presence of vasculopathy [5, 6, 10], although it was possible that by the time autopsy was performed, the cellular reaction might have gone leaving evidence only of endothelial cell injury [2]. Nonetheless, perivascular lymphocytic infiltrates were found relatively frequently in NPSLE [2].

Although SLE patients without neuropsychiatric manifestations showed several vascular changes, such as microinfarction, macroinfarction and vasculitis, they were more frequently observed in NPSLE patients [10]. By contrast, microthrombi were found exclusively in NPSLE patients [10]. As for the distribution of vasculopathy, diffuse vasculopathy was much more common in NPSLE, whereas the incidence of focal vasculopathy was comparable between SLE alone and NPSLE (40–60%) [10].

On microscopic observation, there are gliosis, necrosis, focal edema, inflammatory infiltrates, demyelination, cytotoxic edema and chronic hypoperfusion at WMHIs on MRI scans [11, 12]. GMHIs are usually larger in size than WMHIs and are associated with cerebrovascular disease and seizures [13], where vasogenic edema and direct neuronal autoantibody-mediated damage are observed, as is the case with paraneoplastic syndromes [14–17].

In general population, atrophy is usually considered as part of physiologic phenomenon due to ageing. It was found that atrophy had higher prevalence in patients with SLE irrespective of the presence of neuropsychiatric involvement than in normal population [18]. Focal gyrus atrophy, which is usually associated with WMHIs or GMHIs, may reflect local hypoperfusion due to chronic microangiopathy. It has been disclosed that the presence of regional atrophy of corpus callosum and hippocampus is associated with cognitive dysfunctions in NPSLE [19, 20].

A review of the data on autopsy and 15-Oxygen brain scan studies shows that diffuse vascular abnormalities occur in the brainstem of most patients with NPSLE [2]. Although these abnormalities in small vessels result in microscopic infarcts and hemorrhages in many cases, they are not always associated with severe ischemia enough to cause infarction [2]. Rather, abnormalities of cerebral vasculature and of cerebral metabolism were found to be reversible in some patients [2].

4.3 The Pathogenesis of Vasculopathy in NPSLE

It has been well appreciated that aPL, including anti-cardiolipin antibodies and lupus anticoagulant, are associated with arterial and venous thrombosis, thrombocytopenia and recurrent spontaneous abortion [21, 22]. In fact, elevated levels of these antibodies have been also found in several patients with cerebrovascular diseases in NPSLE [23].

Previous studies revealed that aPL bind negatively charged phospholipids, some of which may be present on endothelial cell membranes [24]. Thus, such binding of aPL with phospholipids on endothelial cells might result in decreased arachidonic acid release [25]. Subsequently, decreased levels of prostacyclin and increased platelet aggregation might cause thrombosis. It is also possible that aPL may directly influence the functions of endothelial cells without causing thrombosis. Moreover, there are autoantibodies capable of binding to endothelial cells other than aPL.

In a rare occasion, thrombosis mediated by aPL was accompanied by vasculitis (Fig. 4.1) [26]. Although vasculitis in the brain is rather uncommon in SLE, it might be detected in about 10% of patients with NPSLE [27]. Overall, cerebral vasculitis is related with SLE disease activity including lower complement levels and responds to steroid treatment to be reversed [14]. It is therefore suggested that the cerebral thrombosis plus arteritis in the patient in Fig. 4.1 might have been caused by a type of vasculitis syndrome due to SLE, but not by aPL. In general, cerebral vasculitis in SLE might be diffuse or localized to one region of one vessel, affecting small or large arteries [28].

Microvasculopathy in SLE was attributed to deposition of immune complexes [2]. Notably, it has been revealed that activation of complement appears to play an important role in the development of microvasculopathy in NPSLE [29]. Consistently, recent studies have demonstrated that microthrombi, which were found uniquely in the brain from patients with NPSLE, were associated with C4d and C5b-9 deposits [10]. It is thus suggested that classical complement activation initiated by immune complexes might be the link between autoantibody mediated inflammation and thromboischemic injury (microthrombi) in NPSLE [10].

It should be also noted that autoantibodies, including aPL, causing direct injury to endothelial cells would be involved in the deposition of complements. Thus, accumulation of antibodies in small vessels most likely leads to activation of the classical complement pathway, endothelial injury and the subsequent formation of microthrombi in anti-phospholipid syndrome (APS) [30] and in thrombotic

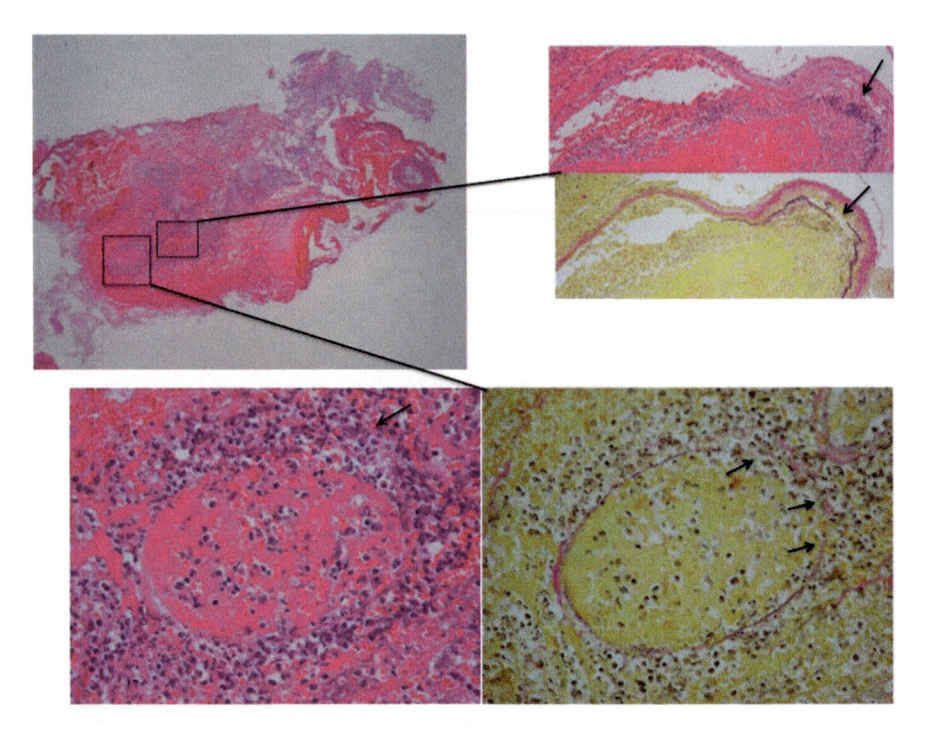

Fig. 4.1 Thrombo-endarteritis in the brain tissue from a patient with diffuse NPSLE and antiphospholipid syndrome (APS), showing brain hematoma. Ajacent to the hematoma, there is necrotizing vasculature fully filled with thrombi and encircled by inflammatory cells, indicating the disrupted lamina elastica interna (arrows) (HE stain and EVG stain)

microangiopathy [31]. On the other hand, it is likely that continuous exposures of the cerebral endothelium to autoantibodies may result in complement activation and endothelial injury in basically all SLE patients irrespective of the presence of overt NPSLE [32]. Therefore, a second hit by other factors, such as infection, pregnancy and drug toxicity [33, 34], is required for the blood-brain barrier (BBB) damages with the formation of C4d- and C5b-9-associated microthrombi, leading to the development of overt NPSLE [10].

4.4 Diffuse Psychiatric/Neuropsychological Syndromes

4.4.1 Acute Confusional State

Acute confusional state (ACS) is the most severe manifestation in diffuse psychiatric/neuropsychiatric syndrome. Previous studies revealed that about 50% of the patients show abnormalities on MRI scans, while the other patients present normal MRI

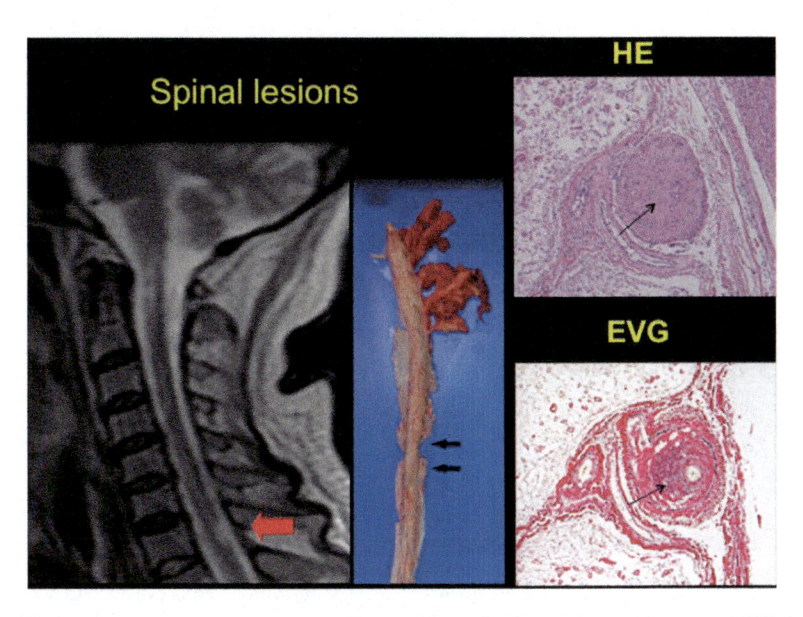

Fig. 4.2 Spinal cord lesions in a patient with myelitis. Left: T2 weightened image on MRI shows lower cervical lesion (red arrow). Middle: Spinal cord lesion at autopsy (black arrows). Right: Thickening of the wall along with mononuclear cell infiltration (upper arrow) (HE stain) in spinal cord arteries with disruption of elastic lamina (lower arrow) (EVG stain)

scans [35]. Interestingly, the presence of abnormalities on MRI scans is associated with a poor prognosis [35]. The pathological features of such abnormal density area have not been reported. However, we disclosed the presence of vasculitis in the spinal cord of a patient who presented transverse myelitis extended to disseminated encephalitis with ACS [36]. In this patient, autopsy findings confirmed the presence of liquefaction necrosis in the entire circumference of the whole spinal cord along with intimal hyperplasia and obliteration of the small arteries, accompanied by mononuclear cell infiltration and disruption of internal elastic lamina (Fig. 4.2). It is most likely that this patient developed longitudinal transverse myelitis through spinal cord vasculitis, which extended to brainstem and brain parenchyma, leading to the development of ACS (Fig. 4.3) [36]. Notably, abnormal high density areas on FLAIR images of MRI scans ameliorated after high doses of steroids [36]. It is therefore suggested that reversible abnormal high density areas might be caused by vasculitis in this patient.

By contrast, the other 50% of patients with ACS did not show any abnormalities on MRI scans [35]. Notably, we also previously described an SLE patient with ACS who died of acute pulmonary hemorrhage [37]. At autopsy, macroscopic findings showed only microinfarction in the brain along with microthrombi (Fig. 4.4 Left, Middle). However, the expression of IL-6 mRNA in neurons in granular layers of hippocampus was evident in this patient (Fig. 4.4 Right) [37]. It is assumed that the influx of neuron-reactive autoantibodies might result in the alteration of the function of neurons without affecting their morphology.

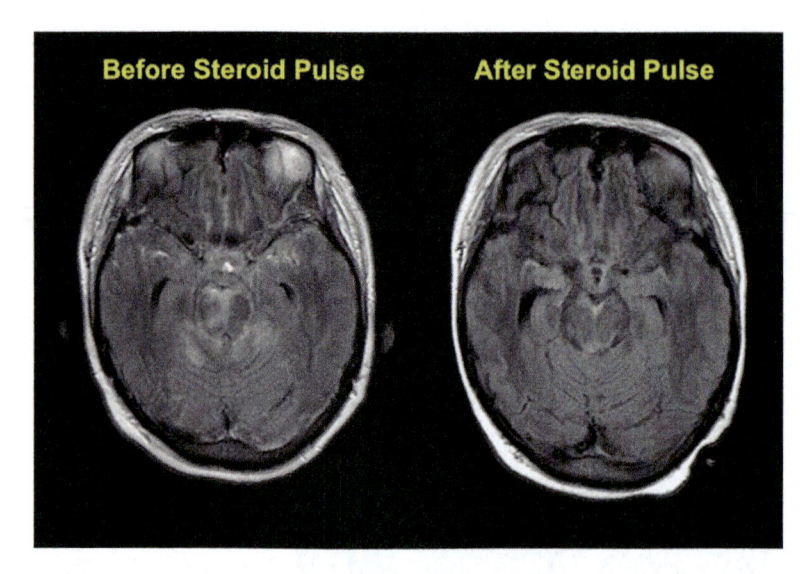

Fig. 4.3 FLAIR images on MRI of the brainstem in a patient with myelitis extended to brainstem encephalitis. After the steroid pulse therapy, high intensity lesions almost completely disappeared (the same patient as in Fig. 4.2)

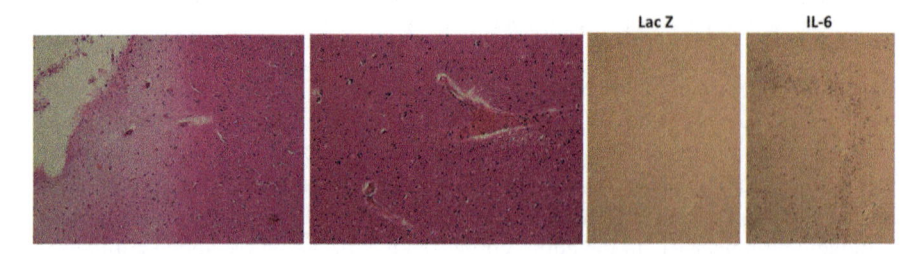

Fig. 4.4 Histopathology of the hippocampus in a patient with acute confusional state (ACS) who died of pulmonary hemorrhage. Left: Microinfarction (HE stain), Middle: Scattered microthrombi (HE stain), Right: Demonstration of IL-6 mRNA expression in granular neurons in the hippocampus by in situ hybridization

It has now been well appreciated that the integrity of the BBB is very important in the pathogenesis of SLE related neuropathology [38]. Processes leading to brain dysfunction in SLE probably involve abnormal endothelial-white blood cell interactions that allow proteins or cells access to the CNS. In fact, it has been demonstrated that the BBB damages in ACS were more severe than those in non-ACS diffuse NPSLE [39, 40]. However, the relationship between BBB damages and brain MRI abnormalities has not been explored [35]. Of note, recent studies disclosed that high field MRI was unable to detect most microvascular and thromboischaemic injuries that were visible on histopathological examinations

[10]. By contrast, white matter lesions may be observed in SLE patients who do not have NP symptoms as well as in normal population [41]. Taken together, it is likely that the microvascular and thromboischaemic injuries, which cannot detected by MRI, might well cause severe BBB damages that might allow entry of high amounts of neuron-reactive autoantibodies, such as anti-NR2 [39]. Of course, it should be noted that brain MRI abnormalities might indicate the presence of vasculitides which affect the prognosis of the patients in ACS [35].

4.4.2 Non-ACS Diffuse NPSLE

As for non-ACS manifestations of diffuse NPSLE, pathological features may include small vessel non-inflammatory vasculopathy, microvessel occlusion, multifocal microinfarcts, microhaemorrhages and cortical atrophy of various degree. Notably, although the presence of aPL is associated with focal NPSLE, including cerebrovascular disease and seizure disorders, they are also frequently present in patients with non-ACS diffuse NPSLE manifestations such as cognitive dysfunction [42–44]. Thus, in the longitudinal studies designed to examine the relationship between serum aPL levels and cognitive dysfunction in SLE patients, it has been demonstrated that the persistent positive aPL is correlated with cognitive dysfunction [43, 45, 46]. Consistently, the results of studies with volumetric magnetic transfer imaging (MTI), detecting tissue damages not visible on conventional MRI, also provided evidence that, apart from macroscopic cerebral infarctions (cerebrovascular diseases), aPL may play a role in the pathogenesis of diffuse microscopic brain damage resulting in cognitive dysfunction in NPSLE [47].

Anti-NMDA receptor NR2 antibodies (anti-NR2) may also play a role in non-ACS of diffuse NPSLE, including cognitive dysfunction and other psychiatric manifestations. Although some cross-sectional studies found no relationship between anti-NR2 and any clinical manifestations [48], others have disclosed significant correlation between anti-NR2 and both cognitive dysfunction and depression [49, 50]. It is possible that the magnitude and degree of BBB dysfunction along with the type and level of autoantibodies may be the determining factor regarding their pathogenicity in the brain. Accordingly, we have demonstrated that the severity of BBB damages in ACS was higher compared with that in non-ACS diffuse NPSLE [39, 40]. As for the relationship between anti-NR2 and brain pathology, it has been demonstrated that the positive anti-NR2 in CSF was significantly associated with hippocampal atrophy [51]. Since anti-NR2 have been demonstrated to react with endothelial cells to induce the production of inflammatory cytokines [52], it is possible that these antibodies might also result in the microvascular changes in the brain.

4.5 Neurologic Syndromes

4.5.1 Cerebrovascular Disease and Reversible Focal Neurological Deficits

Among various manifestations in neurologic syndromes, cerebrovascular disease (CVD) is defined as neurologic deficits due to arterial insufficiency or occlusion, venous occlusion, or hemorrhage [53]. Notably, APS cause thrombosis, which was the most common cause of death in the cohort of the European Working Party on 1000 patients with SLE [54]. The most common thrombotic events in these patients were CVD (11.8%), followed by myocardial infarction (7.4%) and pulmonary embolism (5.9%) [54].

Most of patients with CVD suffer from irreversible neurological damages, which do not respond to steroid therapy [55]. On the other hand, there are accumulating reports on reversible focal neurological deficits in SLE patients, which respond to steroid therapy [56], although they are not included in the 1999 ACR nomenclature [53]. Such focal lesions might occur anywhere in the brain, including cerebral white and gray matters [56] and brainstem (Fig. 4.5). It is considered that the reversible lesions in these patients might be caused by cerebral vasculitis because of the reversibility by steroid therapy as well as the elevation of CSF IL-6 [56]. Thus, "reversible focal neurological deficits" should be added as an independent clinical entity to the 1999 ACR nomenclature.

4.5.2 Seizures

In general, seizures in NPSLE consist of 2 distinct types. One type of seizures might result from structural lesions arising from a wide range of vascular abnormalities, such as microinfarcts, cortical atrophy, macroinfarcts, and hemorrhage [2, 32]. Again, similar to CVD, aPL (anti-cardiolipin antibodies, anti-β2GP1 antibodies, lupus anticoagulant) were associated with seizures [57–59]. The other type of seizures arises in consequence of diffuse psychiatric/neuropsychological syndromes, especially ACS, in which neuron-reactive autoantibodies may play a more direct role [39, 40, 60]. In this type of seizures, structural lesions would probably not be demonstrable.

4.5.3 Cranial Neuropathy

Cranial neuropathy has been categorized in peripheral nervous system involvement in the 1999 ACR nomenclature [53]. However, it may be caused by brain parenchymal lesions. Thus, similarly to visual defects, cranial neuropathy may result from lesions

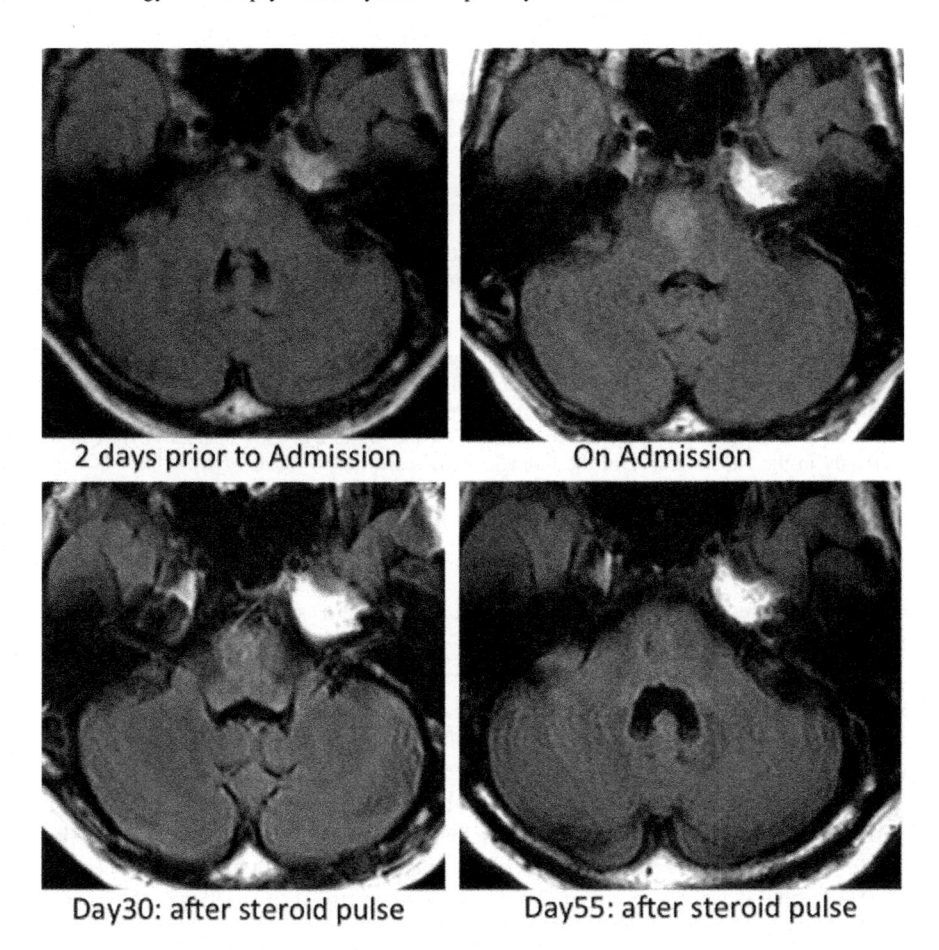

4 Pathology of Neuropsychiatric Systemic Lupus Erythematosus 51

Fig. 4.5 FLAIR images on MRI of a patient with brainstem lesion. Two days prior to admission, no symptoms were observed despite the brainstem lesion on MRI. After the steroid therapy, high intensity lesion on MRI disappeared along with the recovery from left hemiparesis

at any levels of the motor cortex, corticobulbar tracts, brainstem, or peripheral nerves [6]. It should be noted that most cranial nerve deficits were associated with microscopic infarcts in the brainstem or cerebral cortex [4, 6].

4.5.4 Myelopathy

In the 1999 ACR criteria, myelopathy does not discriminate neuromyelitis optica spectrum disorder (NMOSD) caused by anti-aquaporin 4 (AQP4) antibodies from transverse myelitis due to vasculitis. These 2 conditions result in different prognosis,

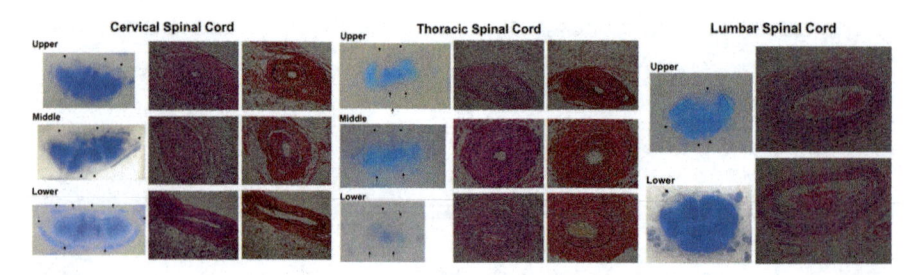

Fig. 4.6 Histopathology of the spinal cord on autopsy in a patient with myelitis extended to disseminated brainstem encephalitis. Weak staining on Kluver-Barrera staining is noted in whole lower cervical spinal cord, throacic spinal cord and upper lumber spinal cord. Intimal thickening, mononuclear cell infiltration and disruption of internal elastica lamina are noted in arteries through cerrvical to lumbar areas. Liquefaction necrosis on the white and grey matters is seen most markedly in the circumference of the thoracic spinal cord (arrow heads). Left: Kluver-Barrera staining ($\times$ 1), Middle: HE staining ($\times$ 50), Right: EVG staining ($\times$ 50)

and therefore need to be differentiated from each other as soon as possible [53]. On the other hand, it has been reported that myelopathy in SLE can be categorized into 2 types [61]. One is myelitis with white matter involvement (upper-motor neuron syndrome with spasticity and hyperreflexia), while the other is that with gray matter signs (lower motor neuron syndrome with flaccidity and hyporeflexia). Patients in the former group were more likely to meet criteria for NMOSD with positive NMO-IgG (anti-AQP4 antibodies) and were also more likely to have aPL (lupus anticoagulant) [61]. Since regional ischemia has been noted to trigger the up-regulation of aquaporin 4 [62], it is likely that aPL may result in the up-regulation of AQP4 antigen through local ischemia, leading to NMO pattern of disease mediated by NMO-IgG [61].

On the other hand, patients in the latter group more likely have irreversible paraplegia with a cerebrospinal fluid profile indistinguishable from bacterial meningitis, presenting with prodromes of fever and urinary retention [61]. We have previously disclosed the histopathological findings in a patient with myelitis with gray matter signs [36]. As shown in Fig. 4.6, autopsy findings confirmed the presence of liquefaction necrosis in the entire circumference of the whole spinal cord along with intimal hyperplasia and obliteration of the small arteries, accompanied by mononuclear cell infiltration and disruption of internal elastic lamina, confirming that the longitudinal transverse myelitis resulted from spinal cord vasculitis [36]. It should be noted that the intensity of tissue destruction was the strongest at mid-thoracic regions (Th3–8), the vasculitic changes were observed throughout the whole spinal cord [36]. Since anti-NR2 were positive in this patient, it is possible that these antibodies might be involved in the pathogenesis of vascltitis leading to myelitis [36, 52]. Moreover, it is evident that this patient presented severe white matter demyelination along with gray matter necrosis. It is assumed that gray matter is more vulnerable to ischemic changes due to vasculitis since it is more distant from the vessels. However, both white matter and gray matter can be affected when the severe ischemia takes place.

4.5.5 Peripheral Neuropathy

Abnormalities of vessels supplying the peripheral nerve might result in mononeuritis multiplex in a manner similar to polyarteritis nodosa [4, 6]. By contrast, demyelination in the absence of demonstrable vasculopathy has been described frequently [6]. In the recent study by Oomatia A et al., the overall prevalence of peripheral neuropathies was 5.9%, with 66.7% having peripheral neuropathies attributable to SLE [63]. It was disclosed that 17.1% of the patients with SLE-related peripheral neuropathies had biopsy-proven small-fiber neuropathies. Thus, small-fiber neuropathies were much more common than acute inflammatory demyelinating neuropathies (Guillain-Barre's syndrome), plexopathies and mononeuritis multiplex, which was seen in only 7.3% of the patients with peripheral neuropathy (6 of 82 patients) [63]. Notably, the skin biopsy findings in small-fiber neuropathy patients suggested that distinct mechanisms might target the dorsal root ganglia as well as distal axons [63].

4.6 Other Pathological Features in NPSLE

4.6.1 Choroid Plexus

Considerable interest was generated by the demonstration of deposition of diffuse immunoglobulin, and in some cases, C3 complement in the choroid plexus at autopsy of patients with NPSLE [2]. Such deposits have also been demonstrated in the choroid plexus of old NZB/WF1 mice and in other experimental animal models. Although many investigators tried to explore the significance of immune-complexes deposition in the choroid plexus in the pathogenesis of NPSLE, its role declined due to the lack of its specificity in NPSLE [2].

4.6.2 Microglia

Recent studies have shed light on the roles of microglia in the pathogenesis of NPSLE. Thus, type I IFN stimulates microglia to become reactive and engulf neuronal and synaptic materials, resulting in synapse loss and behavioral phenotypes in lupus-prone mice [64]. In addition, increased type I IFN signaling in postmortem hippocampal brain sections from patients with SLE has been also demonstrated [64].

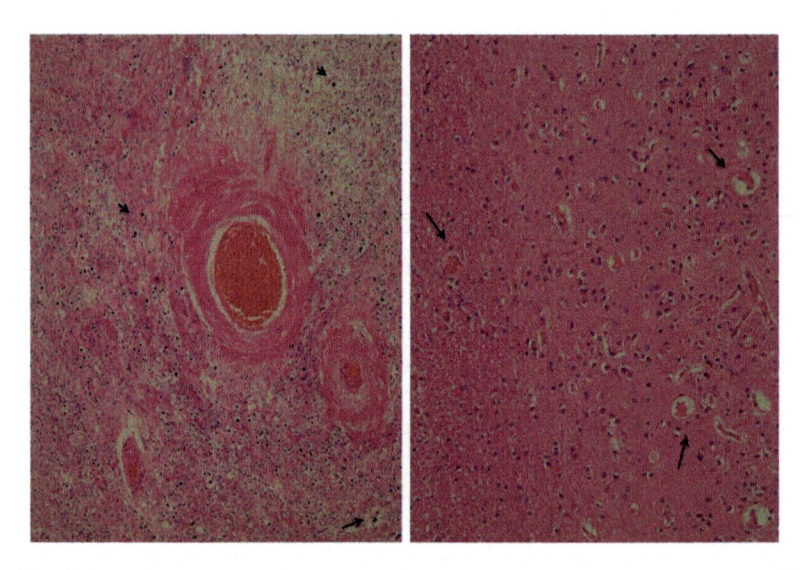

Fig. 4.7 Histopathology of the brain on autopsy in a patient with acute confusional state (ACS) of diffuse NPSLE. Left: Perivenous lesion consisting of foci of coagulation necrosis surrounding the veins along with scattered calcification (arrows). Right: Nearby the perivenous lesion, scattered of microthrombi were observed (arrows). HE staining

On the other hand, recent studies have also demonstrated that the sera from an SLE patient caused morphological changes in the microglia, with an increase in expression of MHC class II antigen and CD86 as well as an enhanced release of nitric oxide and proinflammatory cytokines, thus underscoring the potential role of microglia in NPSLE [65].

4.6.3 Perivenous Changes

At autopsy of an SLE patient with ACS, multiple perivenous, well-demarcated foci of brownish discoloration were seen scattered throughout the cerebral white matter and basal ganglia [66]. Histopathologically these lesions consisted of foci of coagulation necrosis surrounding the veins (Fig. 4.7 Left). There are also scattered calcification. The veins in the foci showed fibrous thickening of the walls with no evidence of vasculitis [66]. Of note, around the lesion of coagulation necrosis scattered microthrombi were found (Fig. 4.7 Right).

Similar pathological changes have been shown in the other studies. Mizutani T et al. [67] described the multiple demyelinated and necrotic foci associated with perivenous cell infiltration and the exudation of fibrinoid material. Matsumoto R et al. [68] reported perivenous lesions composing of perivenous spongy changes unaccompanied by an inflammatory cell infiltration, but with severe disruption of both myelin sheaths and axons. The significance of perivenous changes above mentioned needs to be explored in accumulating cases in the future, although it is likely that leakages of some toxic substances from the vein might result in the tissue damages.

4.7 Summary

Non-inflammatory vasculopathy in small vessels, such as microvessel occlusion, multifocal microinfarcts, microhaemorrhages and cortical atrophy were most common features observed in NPSLE at autopsy. True arteritis, characterized by cellular infiltration of vessel walls, was noted only in a fraction of the patients. It is suggested that reversible abnormal high density areas, which improve in response to steroid treatment, possibly caused by vasculitis, can be seen occasionally in SLE patients. At autopsy of diffuse NPSLE without MRI abnormalities, macroscopic findings showed only micro-infarction in the brain, where neuron-reactive autoantibodies might affect the functions of neurons directly. Generally, a variety of autoantibodies are involved in vascular lesions, including aPL, anti-NR2 and anti-ribosomal P, anti-Sm and anti-RNP. Neurological manifestations such as stroke, cortical visual defects, cranial nerve abnormalities, seizures may be associated with structural lesions secondary to vasculopathy, possibly caused by aPL. Foci of coagulation necrosis surrounding the veins can be observed in patients with diffuse NPSLE, although the significance of these finding need to be explored. The roles of microglia in the pathogenesis of NPSLE have been recently demonstrated.

References

1. Gibson T, Myers AR. Nervous system involvement in systemic lupus erythematosus. Ann Rheum Dis. 1975;35:398–406.
2. Harris EN, Hughes GR. Cerebral disease in systemic lupus erythematosus. Springer Semin Immunopathol. 1985;8:251–66.
3. Govoni M, et al. The diagnosis and clinical management of the neuropsychiatric manifestations of lupus. J Autoimmun. 2016;4:41–72.
4. Arinuma Y, et al. Brain magnetic resonance imaging in patients with diffuse psychiatric/neuropsychological syndromes in systemic lupus erythematosus. Lupus Sci Med. 2014;1:e000050.
5. Ellis SG, Verity MA. Central venous system involvement in systemic lupus erythematosus: a review of neuropathologic findings in 57 cases, 1955-1977. Semin Arthritis Rheum. 1979;8:212–21.
6. Johnson RT, Richardson EP. The neurological manifestations of systemic lupus erythematosus: a clinical-pathological study of 24 cases and review of the literature. Medicine. 1968;47:337–69.
7. Hanly JG, et al. Brain pathology in systemic lupus erythematosus. J Rheumatol. 1992;19:732–41.
8. Brooks WM, et al. The histopathologic associates of neurometabolite abnormalities in fatal neuropsychiatric systemic lupus erythematosus. Arthritis Rheum. 2010;62:2055–63.

9. Ellison D, et al. Intramural platelet deposition in cerebral vasculopathy of systemic lupus erythematosus. J Clin Pathol. 1993;46:37–40.
10. Cohen D, et al. Brain histopathology in patients with systemic lupus erythematosus: identification of lesions associated with clinical neuropsychiatric lupus syndromes and the role of complement. Rheumatology. 2017;55:77–86.
11. Sibbitt WL, et al. Magnetic resonance imaging and brain histopathology in neuropsychiatric systemic lupus erythematosus. Semin Arthritis Rheum. 2010;40:32–52.
12. Sibbitt WL, et al. Fluid attenuated inversion recovery (FLAIR) imaging in neuropsychiatric systemic lupus erythematosus. J Rheumatol. 2003;30:1983–9.
13. Jeong HW, et al. Brain MRI in neuropsychiatric lupus: associations with the 1999 ACR case definitions. Rheumatol Int. 2015;35:861–9.
14. Sibbitt WL, et al. Neuroimaging in neuropsychiatric systemic lupus erythematosus. Arthritis Rheum. 1999;42:2026–38.
15. Emmer BJ, et al. Selective involvement of the amygdala in systemic lupus erythematosus. PLoS Med. 2006;3:e499.
16. Dalmau J, Rosenfeld MR. Paraneoplastic syndromes of the CNS. Lancet Neurol. 2008;7:327–40.
17. Hoeftberger R, et al. Update on neurological paraneoplastic syndromes. Curr Opin Oncol. 2015;27:489–95.
18. Kozora E, et al. Magnetic resonance imaging abnormalities and cognitive deficits in systemic lupus erythematosus patients without overt central nervous system disease. Arthritis Rheum. 1998;41:41–7.
19. Appenzeller S, et al. Cerebral and corpus callosum atrophy in systemic lupus erythematosus. Arthritis Rheum. 2005;52:2783–9.
20. Appenzeller S, et al. Hippocampal atrophy in systemic lupus erythematosus. Ann Rheum Dis. 2006;65:1585–9.
21. Boey ML, et al. Thrombosis in SLE: striking association with the presence of circulating "lupus anticoagulant". Br Med J. 1983;287:1021–3.
22. Harris EN, et al. Anticardiolipin antibodies: detection by radioimmunoassay and association with thrombosis in systemic lupus erythematosus. Lancet. 1983;2:1211–4.
23. Harris EN, et al. Cerebral infarction in systemic lupus: association with anticardiolipin antibodies. Clin Exp Rheumatol. 1984;1:47–51.
24. Harris EN, et al. Cross-reactivity of antiphospholipid antibodies. J Clin Lab Immunol. 1985;16:1–6.
25. Carreras LO, Vermelyn JG. "Lupus" anticoagulant and thrombosis-possible role of l inhibition of prostacyclin formation. Thromb Haemost. 1982;48:38–40.
26. Arinuma Y, et al. Histopathological analysis of cerebral hemorrhage in systemic lupus erythematosus complicated with antiphospholipid syndrome. Mod Rheumatol. 2011;21:509–13.
27. Wallace DJ, Metzger AL. Systemic lupus erythematosus and the nervous system. In: Wallace DJ, Hahn BH, editors. Dubois' lupus erythematosus. 4th ed. Philadelphia: Lea & Febiger; 1993. p. 370–85.
28. Jennekens FGI, Kater L. The central nervous system in systemic lupus erythematosus. Part 2. Pathogenetic mechanisms of clinical syndromes: a literature investigation. Rheumatology. 2002;41:619–30.
29. Belmont HM, et al. Pathology and pathogenesis of vascular injury in systemic lupus erythematosus. Interactions of inflammatory cells and activated endothelium. Arthritis Rheum. 1996;39:9–22.
30. Cohen D, et al. Classical complement activation as a footprint for murine and human antiphospholipid antibody-induced fetal loss. J Pathol. 2011;225:502–11.
31. Chua JS, et al. Complement factor C4d is a common denominator in thrombotic microangiopathy. J Am Soc Nephrol. 2015;26:2239–47.
32. Muscal E, Brey RL. Neurological manifestations of systemic lupus erythematosus in children and adults. Neurol Clin. 2010;28:61–73.

33. Salmon JE, et al. Complement activation as a mediator of antiphospholipid antibody induced-pregnancy loss and thrombosis. Ann Rheum Dis. 2002;61(Suppl 2):ii46–50.
34. Noris M, Remuzzi G. Atypical hemolytic-uremic syndrome. N Engl J Med. 2009;361:1676–87.
35. Abe G, et al. Brain MRI in patients with acute confusional state of diffuse psychiatric/neuropsychological syndromes in systemic lupus erythematosus. Mod Rheumatol. 2017;27:278–83.
36. Tono T, et al. Transverse myelitis extended to disseminated encephalitis in systemic lupus erythematosus: histological evidence for vasculitis. Mod Rheumatol. 2016;26:958–62.
37. Hirohata S, Hayakawa K. Enhanced interleukin-6 messenger RNA expression by neuronal cells in a patient with neuropsychiatric systemic lupus erythematosus. Arthritis Rheum. 1999;42:2729–30.
38. Abbott NJ, et al. The blood-brain barrier in systemic lupus erythematosus. Lupus. 2003;12:908–15.
39. Hirohata S, et al. Blood-brain barrier damages and intrathecal synthesis of anti-N-methyl-D-aspartate receptor NR2 antibodies in diffuse psychiatric/neuropsychological syndromes in systemic lupus erythematosus. Arthritis Res Ther. 2014;16:R77.
40. Hirohata S, et al. Association of cerebrospinal fluid anti-Sm antibodies with acute confusional state in systemic lupus erythematosus. Arthritis Res Ther. 2014;16:450.
41. Kent DL, et al. The clinical efficacy of magnetic resonance imaging in neuroimaging. Ann Intern Med. 1994;120:856–71.
42. Brey RL. Antiphospholipid antibodies in young adults with stroke. J Thromb Thrombolysis. 2005;20:105–12.
43. Hanly JG, et al. A prospective analysis of cognitive function and anticardiolipin antibodies in systemic lupus erythematosus. Arthritis Rheum. 1999;42:728–34.
44. Mikdashi J, Handwerger B. Predictors of neuropsychiatric damage in systemic lupus erythematosus: data from the Maryland lupus cohort. Rheumatology. 2004;43:1555–60.
45. McLaurin EY, et al. Predictors of cognitive dysfunction in patients with systemic lupus erythematosus. Neurology. 2005;64:297–303.
46. Menon S, et al. A longitudinal study of anticardiolipin antibody levels and cognitive functioning in systemic lupus erythematosus. Arthritis Rheum. 1999;42:735–41.
47. Steens SC, et al. Association between microscopic brain damage as indicated by magnetization transfer imaging and anticardiolipin antibodies in neuropsychiatric lupus. Arthritis Res Ther. 2006;8:R38.
48. Lapteva L, et al. Anti-N-methyl-D-aspartate receptor antibodies, cognitive dysfunction, and depression in systemic lupus erythematosus. Arthritis Rheum. 2006;54:2505–14.
49. Omdal R, et al. Neuropsychiatric disturbances in SLE are associated with antibodies against NMDA receptors. Eur J Neurol. 2005;12:392–8.
50. Arinuma Y, et al. Association of cerebrospinal fluid anti-NR2 glutamate receptor antibodies with diffuse neuropsychiatric systemic lupus erythematosus. Arthritis Rheum. 2008;58:1130–5.
51. Lauvsnes MB, et al. Association of hippocampal atrophy with cerebrospinal fluid antibodies against the NR2 subtype of the N-methyl-D-aspartate receptor in patients with systemic lupus erythematosus and patients with primary Sjøgren's syndrome. Arthritis Rheumatol. 2014;66:3387–94.
52. Yoshio T, et al. IgG anti-NR2 glutamate receptor autoantibodies from patients with systemic lupus erythematosus activate endothelial cells. Arthritis Rheum. 2013;65:457–63.
53. ACR Ad Hoc Committee on Neuropsychiatric Lupus Nomenclature. The American College of Rheumatology nomenclature and case definitions for neuropsychiatric lupus syndromes. Arthritis Rheum. 1999;42:599–608.
54. Cervera R, et al. Morbidity and mortality in systemic lupus erythematosus during a 10-year period: a comparison of early and late manifestations in a cohort of 1,000 patients. Medicine. 2003;82:299–308.
55. Toubi E, et al. Association of antiphospholipid antibodies with central nervous system disease in systemic lupus erythematosus. Am J Med. 1995;99:397–401.

56. Kimura M, et al. Reversible focal neurological deficits in systemic lupus erythematosus: report of 2 cases and review of the literature. J Neurol Sci. 2008;272:71–6.
57. Sanna G, et al. Neuropsychiatric manifestations in systemic lupus erythematosus: prevalence and association with antiphospholipid antibodies. J Rheumatol. 2003;30:985–92.
58. Andrade RM, et al., LUMINA Study Group. Seizures in patients with systemic lupus erythematosus: data from LUMINA, a multiethnic cohort (LUMINA LIV). Ann Rheum Dis. 2008;67:829–34.
59. Appenzeller S, et al. Epileptic seizures in systemic lupus erythematosus. Neurology. 2004;63:1808–12.
60. Bluestein HG, et al. Cerebrospinal fluid antibodies to neuronal cells: association with neuropsychiatric manifestations of systemic lupus erythematosus. Am J Med. 1981;70:240–6.
61. Birnbaum J, et al. Distinct subtypes of myelitis in systemic lupus erythematosus. Arthritis Rheum. 2009;60:3378–87.
62. Taniguchi M, et al. Induction of aquaporin-4 water channel mRNA after focal cerebral ischemia in rat. Brain Res Mol Brain Res. 2000;78:131–7.
63. Oomatia A, et al. Peripheral neuropathies in systemic lupus erythematosus: clinical features, disease associations, and immunologic characteristics evaluated over a twenty-five-year study period. Arthritis Rheumatol. 2014;66:1000–9.
64. Bialas AR, et al. Microglia-dependent synapse loss in type I interferon-mediated lupus. Nature. 2017;546:539–43.
65. Wang J, et al. Microglia activation induced by serum of SLE patients. J Neuroimmunol. 2017;310:135–42.
66. Shintaku M, Matsumoto R. Disseminated perivenous necrotizing encephalomyelitis in systemic lupus erythematosus: report of an autopsy case. Acta Neuropathol. 1998;95:313–7.
67. Mizutani T, et al. A case of demyelinating encephalomyelitis with some resemblance to collagen disease. J Neurol. 1977;217(1):43–52.
68. Matsumoto R, et al. Systemic lupus erythematosus with multiple perivascular spongy changes in the cerebral deep structures, midbrain and cerebellar white matter: a case report. J Neurol Sci. 1997;145:147–53.

Chapter 5
Clinical Features

Yoshiyuki Arinuma and Shunsei Hirohata

Abstract Clinical features in neuropsychiatric systemic lupus erythematosus (NPSLE) include a variety of manifestations, which are too complicated to understand. In 1999, the American college of rheumatology (ACR) proposed nomenclature and case definitions of NPSLE as nosology for clinical descriptions and research. Clinically, NPSLE is classified into 2 different categories; one is diffuse psychiatric/neuropsychological syndromes and the other is neurologic syndromes. This discrimination is not always advantageous for clinical practice including a diagnosis and therapeutic intervention. Moreover, the severities of each manifestation differ among NPSLE. For example, acute confusional state is very severe form with poor prognosis, whereas headache, mood disorders, anxiety disorders and cognitive dysfunction are sometimes mild, and are also common in individuals without SLE. Also, we need to remember that NPSLE can be developed even in the absence of systemic disease activities. Finally, some neuropsychiatric manifestations have not been adequately defined in the ACR nomenclature.

Keywords NPSLE · ACR nomenclature · Diffuse psychiatric/neuropsychological Syndromes · Acute confusional state · Neurologic syndromes

---

Electronic supplementary material: The online version of this chapter at https://doi.org/10.1007/978-3-319-76496-2_5. contains supplementary material, which is available to authorized users.

Y. Arinuma (✉)
Department of Rheumatology and Infectious diseases, Kitasato University School of Medicine, Sagamihara, Kanagawa, Japan

S. Hirohata
Department of Rheumatology
Nobuhara Hospital, Tatsuno, Hyogo, Japan

Department of Rheumatology & Infectious Diseases
Kitasato University School of Medicine, Sagamihara, Kanagawa, Japan

© Springer International Publishing AG, part of Springer Nature 2018
S. Hirohata (ed.), *Neuropsychiatric Systemic Lupus Erythematosus*,
https://doi.org/10.1007/978-3-319-76496-2_5

5.1 Introduction

Neuropsychiatric involvement is rather common in systemic lupus erythematosus (SLE), including the psychiatric syndromes and the neurologic syndromes. For the first time, a case with coma was published in 1875 as an SLE patient with neuropsychiatric syndrome [1].

Neuropsychiatric manifestations occur as frequently as in 56.3% of SLE patients according to the result of meta-analysis [2]. NPSLE includes a variety of clinical symptoms, resulting from damages in central nervous system (CNS) and peripheral nervous system (PNS) as primary or secondary complications. The American college of rheumatology (ACR) established nomenclature and case definitions of neuropsychiatric systemic lupus erythematosus (NPSLE) as nosology for clinical descriptions and research in SLE.

In this chapter, we present the clinical features of NPSLE required for physicians, especially for rheumatologists, explaining definitions and clinical points by respective manifestation based on the ACR 1999 nomenclature. Also, we discuss about other manifestations which have not been adequately described in the ACR case definitions.

5.2 Classification

The ACR 1999 nomenclature defines 19 neuropsychiatric syndromes as NPSLE, including 12 syndromes of CNS and 7 of PNS (Table 5.1). The CNS syndromes of NPSLE were further classified into 2 categories, including diffuse psychiatric/neuropsychological syndromes and neurologic syndromes. Diffuse psychiatric/neuropsychological syndromes consist of 5 psychiatric disorders, including acute confusional state, anxiety disorder, cognitive dysfunction, mood disorder and psychosis. Neurologic syndromes comprise of aseptic meningitis, cerebrovascular disease, demyelinating syndrome, headache, movement disorder, myelopathy and seizure disorders [3]. In this nomenclature, the ACR also emphasized that the classification criteria for NPSLE were intended for purposes of classification and reporting, but not for clinical judgment or for use in making a clinical diagnosis.

ACR also released the guideline including recommendation of the diagnostic tests to specify NPSLE. The complete case definitions for 19 manifestations of NPSLE listed in Table 5.1 were given as Appendix A, including definition, diagnostic criteria, and methods for ascertainment. Comorbid conditions and concomitant factors that may cause identical symptoms and which should be excluded before attributing the syndrome to SLE were also listed ("exclusions"). In some instances, it may not be possible to judge whether neuropsychiatric findings are due to lupus or to other causes, such as irreversible, existing conditions (e.g., diabetes) that cannot be cured or corrected, or drugs that cannot be with-

5 Clinical Features

Table 5.1 Neuropsychiatric manifestations in SLE (1999 ACR)

Central nervous system
<u>Diffuse psychiatric/neuropsychological syndromes</u>
Acute confusional state
Anxiety disorder
Cognitive dysfunction
Mood disorder
Psychosis
<u>Neurologic syndromes</u>
Aseptic meningitis
Cerebrovascular disease
Demyelinating syndrome
Headache (benign intracranial hypertension)
Movement disorder
Myelopathy
Seizure disorders
Peripheral nervous system
<u>Neurologic syndromes</u>
Neuropathy, cranial
Acute inflammatory demyelinating polyradiculoneuropathy (Guillain-Barré syndrome)
Autonomic disorder
Mononeuropathy
Myasthenia gravis
Plexopathy
Polyneuropathy

held, replaced, or withdrawn for the purpose of exclusion (e.g., corticosteroids). These are termed "associations" (as opposed to "exclusions"). In the ascertainment part, objective methods were provided to confirm and evaluate the individual lesion. The ACR nomenclature also included appendix B providing basic laboratory data to be reported and appendix C providing short batteries to assess respective neuropsychiatric manifestation in patients.

5.3 Case Definition and Clinical Significance

5.3.1 Diffuse Psychiatric/Neuropsychological Syndromes

The terminology for each manifestation in diffuse NPSLE has been assigned according to the Diagnostic and Statistical Manual of Mental Disorders, Fourth Edition (DSM-IV).

5.3.1.1 Acute Confusional State

Acute confusional state (ACS) is the severest form of diffuse NPSLE characterized by a fluctuating level of consciousness of acute or subacute onset. Patients sometimes go into a coma (Video 5.1), even if magnetic resonance imaging (MRI) revealed no organic lesions (Fig.5.1). ACS was previously called as acute organic brain syndrome. ACS might be accompanied by disturbances of cognition, mood, affect, and/or behavior. The results of a recent meta-analysis demonstrated that patients with ACS have the significantly poorer prognosis with mortality of hazard ratio 3.4, which was the highest even after multivariate analysis [4].

The definition of ACS from appendix A is "disturbance of consciousness or level of arousal with reduced ability to focus, maintain, or shift attention, which develops over a short period of time (hours to days) and tend to fluctuate during the course of the day". As shown in Table 5.2, for the *diagnosis* of ACS one of the followings is required: A. Acute or subacute change in cognition that may include memory deficit and disorientation, or B. A change in behavior, mood, or affect (e.g., restlessness, overactivity, reversal of the sleep/wakefulness cycle, irritability, apathy, anxiety, mood lability, etc.). It is important to exclude possibilities other than SLE, such as metabolic encephalopathies. The key issue in ACS is the presence of conscious disturbance (delirium) with either cognitive dysfunction and/or mood and behavior dysfunction.

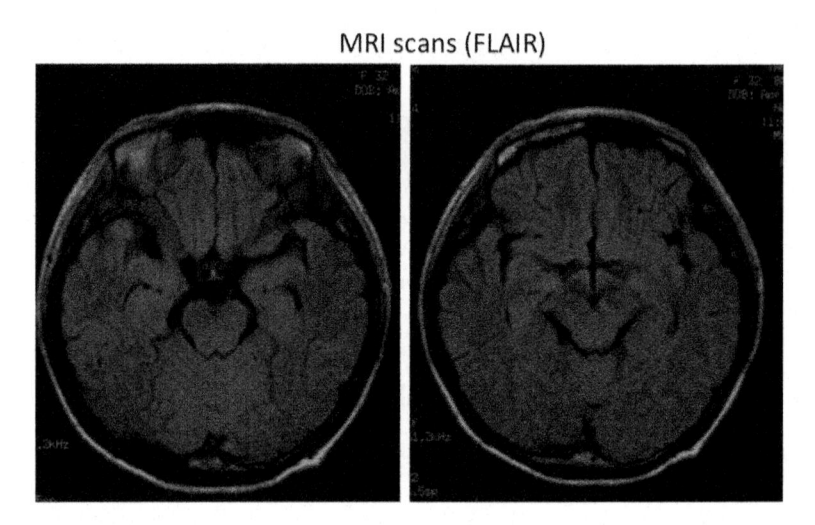

MRI scans (FLAIR)

Fig. 5.1 An SLE patient with acute confusional state. MRI scans revealed no organic lesions

Table 5.2 Diagnostic criteria, exclusions, association, ascertainment, and record of acute confusional state from ACR nomenclature and case definitions in appendix A

Acute confusional state

Diagnostic criteria
- Disturbance of consciousness or level of arousal with reduced ability to focus, maintain, or shift attention, and one or more of the following developing over a short period of time (hours to days) and tending to fluctuate during the course of the day
    A. Acute or subacute change in cognition that may include memory deficit and disorientation
    B. A change in behavior, mood, or affect (e.g., restlessness, overactivity, reversal of the sleep/wakefulness cycle, irritability, apathy, anxiety, mood lability, etc.)

Exclusions
- Primary mental/neurologic disorder not related to SLE
- Metabolic disturbances (including glucose, electrolytes, fluid, osmolarity)
- Substance or drug-induced delirium (including withdrawal)
- Cerebral infections
    NB: Preexisting cognitive deficits are not an exclusion. If acute confusional state is superimposed on preexisting cognitive deficits, diagnose both

Associations
- Marked psychosocial stress
- Corticosteroid use
- Thrombotic thrombocytopenic purpura/hemolytic uremic syndrome

Ascertainment
- Disturbed consciousness: Clinical observation, mental status, and neurologic examination
- Cognitive function: Mental status examination, including instruments such as the mini mental status examination
- Mood and behavioral dysfunction: Clinical observation, history by patient and others, standardized instruments (e.g., hospital anxiety and depression scale)
- Determine from the individual or from informants the impact of disturbance on daily life, previous occupational and social functioning

Record
- Basic descriptors
    NB: Preexisting cognitive deficits are not an exclusion. If acute confusional state is superimposed on preexisting cognitive deficits, diagnose both

5.3.1.2 Anxiety Disorder

Anxiety disorder (AD) is more common in patients with SLE compared to healthy individuals and is sometimes independent of the disease activity of SLE [5]. Higher levels of AD and a younger age may increase the risk of depression in SLE patients [6]. *The definition* of AD in the ACR nomenclature is "anticipation of danger or misfortune accompanied by apprehension, dysphoria, or tension". Among AD, there are generalized anxiety, panic disorder, panic attacks, and obsessive-compulsive disorders. For the *diagnosis* of AD, both of the following should be satisfied: A. Prominent anxiety, panic disorder, panic attacks, or obsessions or compulsions, and B. Disturbance causes clinically significant distress or impaired social, occupational, or other important functioning. It is sometimes necessary to rule out the possibility of adjustment disorder with anxiety or anxiety occurring during the other domains of diffuse NPSLE.

5.3.1.3 Cognitive Dysfunction

Cognitive dysfunction (CD) is the most common manifestation in NPSLE [7, 8], which occurs in more than 50% of SLE patients [9]. Notably, CD is 2 times more prevalent in SLE than in the general population [10]. *The definition* of CD is "significant deficits in any or all of the following cognitive functions: simple or complex attention, reasoning, executive skills (e.g., planning, organizing, sequencing), memory (e.g., learning, recall), visual-spatial processing, language (e.g., verbal fluency), and psychomotor speed". CD implies a decline from a higher level of functioning and ranges from mild impairment to severe dementia. It may or may not impede social, educational, or occupational functioning, depending on the function(s) impaired and the severity of impairment. Subjective complaints of CD are common and may not be objectively verified. Neuropsychological testing should be done in suspected CD for confirmation with a help of a neuropsychologist. *Diagnostic criteria* include: A. Documented impairment in one or more of the cognitive domains, such as simple attention, complex attention, memory (e.g., learning and recall), visual-spatial processing, language (e.g., verbal fluency), reasoning/problem solving, psychomotor speed and executive functions (e.g., planning, organizing, and sequencing). B. The cognitive deficits represent a significant decline from a former level of functioning (if known). C. The cognitive deficits may cause varying degrees of impairment in social, educational, or occupational functioning, depending on the function(s) impaired and the degree of impairment. To confirm the diagnosis, standardized neuropsychological tests are required with estimation of premorbid level of functioning. In addition, it is important to determine the impact of CD on social or occupational functioning of patients.

Of note, the previous study demonstrated that CD in SLE patients might be rather stable, not be so progressive over time [11, 12]. It is also important to remember that a number of conditions other than SLE, such as primary CNS disease or injury, chronic medical illness, medication, psychological or psychiatric disturbance, metabolic disturbance, pain fatigue and sleep disturbance, could develop CD [13].

5.3.1.4 Mood Disorder

Mood disorder (MD) is relatively common in patients with SLE. The prevalence of MD is 6–43% with a wide variety [8, 14–24]. According to the study by Bachen EA et al., the prevalence of various types of MD in female lupus patients was as follows: major depressive disorder (47%), specific phobia (24%), panic disorder (16%), obsessive-compulsive disorder (9%), and bipolar I disorder (6%) [23].

Table 5.3 Diagnostic criteria for mood disorders

I. Major depressive-like episode
One or more major depressive episodes with at least five of the following symptoms, including either A or B or both, during a 2-week period and nearly every day
  A. Depressed mood most of the day, by subjective report or observation made by others
  B. Markedly diminished interest or pleasure in all, or almost all, activities most of the day, by subjective report or observation made by others
  C. 1. Significant weight loss without dieting or weight gain (>5% of body weight in 1 month)
   1. Insomnia or hypersomnia, psychomotor agitation or retardation (observable by others, not merely subjective feeling of restlessness or being slowed down)
   1. Fatigue or loss of energy
   1. Feelings of worthlessness or excessive or inappropriate guilt (may be delusional)
   1. Diminished ability to think or concentrate, or indecisiveness
   1. Recurrent thoughts of death (not just fear of dying), recurrent suicidal ideation without a specific plan, or a suicide attempt or a specific plan for committing suicide

II. Mood disorder with depressive features
All of the following
  A. Prominent and persistent mood disturbance characterized by predominantly depressed mood or markedly diminished interest or pleasure in all, or almost all, activities
  B. Full criteria for major depressive-like episode are not met

III. Mood disorder with manic features
Prominent and persistent mood disturbance characterized by predominantly elevated, expansive, or irritable mood

IV. Mood disorder with mixed features
Prominent and persistent mood disturbance characterized by symptoms of both depression and mania; neither predominates

For all mood disorders, symptoms must cause significant distress or impairment in social, occupational, or other important areas of functioning

The definition of MD is "prominent and persistent disturbance in mood characterized by depressed mood or markedly diminished interest or pleasure in almost all activities or elevated, expansive or irritable mood". *Diagnostic criteria* are established separately for (I) Major depressive-like episode, (II) Mood disorder with depressive features, (III) Mood disorder with manic features, (IV) Mood disorder with mixed features, and for all mood disorders (Table 5.3). It is important to rule out MD from steroid induced psychosis, in which mood changes are common.

5.3.1.5 Psychosis

Psychosis due to SLE is observed in 2% of SLE patients in the retrospective study, which occurred within the first year in 60% of patients after initial onset of SLE [25]. Secondary psychosis caused by corticosteroids administration is more frequently found in SLE patients [26–29]. Recent study demonstrated that the presence of SLE could provide more risk for development of steroid induced psychosis than that of other autoimmune diseases treated with corticosteroids [30]. Therefore, diffuse NPSLE and steroid induced psychosis are not antinomy. In other words, they can appear simultaneously based on different mechanisms.

The definition of psychosis is "severe disturbance in the perception of reality characterized by delusions and/or hallucinations". For *diagnosis* of psychosis, all of the following criteria should be fulfilled: A. At least one of delusions, hallucinations without insight needs to be present, B. The disturbance causes clinical distress or impairment in social, occupational, or other relevant areas of functioning, C. The disturbance does not occur exclusively during the course of a delirium, D. The disturbance is not better accounted for by another mental disorder (e.g., mania). Again, it is important to remember that psychosis of diffuse NPSLE and steroid induced psychosis can occur simultaneously.

5.3.2 Neurologic Syndromes of CNS

Neurologic syndromes of CNS (neurologic NPSLE in CNS) consist of seven manifestations derived from damages of a local or restricted area of the brain and/or spinal cord. It is important to discriminate primary neurological diseases or other diseases developing neurologic syndromes. Hereby, clinical characteristics and case definitions of neurologic NPSLE in CNS in the 1999 ACR case definitions are concisely reviewed and discussed.

5.3.2.1 Aseptic Meningitis

Aseptic meningitis has been documented to be one of the specific features in NPSLE [31]. Meningitis was observed in about 1.6% of the patients with SLE, whereas in 40% of such cases no microorganism could be isolated [32]. Although the etiology of aseptic meningitis has not been clear, some reports indicated such neurological syndromes as transverse myelitis and cerebral vasculitis were associated with aseptic meningitis [33–35]. It should be remembered that nonsteroidal anti-inflammatory drugs and trimethoprim-sulfamethoxazole can be a cause of aseptic meningitis in patients with SLE [36–38].

The definition of aseptic meningitis is "syndrome of fever, headache, and meningeal irritation with cerebrospinal fluid (CSF) pleocytosis, and negative CSF cultures". *Diagnostic criteria* require all of the followings: A. Acute or subacute onset of headache with photophobia, neck stiffness, and fever, B. Signs of meningeal irritation, C. Abnormal CSF. Basically, it is impossible to discriminate viral meningitis from aseptic meningitis due to SLE.

5.3.2.2 Cerebrovascular Disease

Cerebrovascular disease consists of ischemic stroke including transient ischemic syndrome, intracerebral and subarachnoid hemorrhage. Anti-phospholipid syndrome is significantly associated with development of stroke [39, 40]. According to the meta-analysis, individuals with SLE have a two-fold higher risk of ischemic stroke, a three-fold higher risk of intracerebral hemorrhage, and an almost four-fold higher risk of subarachnoid hemorrhage compared to the general population [41]. Several studies have demonstrated that the incidence of stroke in SLE was significantly increased by non-SLE factors such as hypertension [42, 43], dyslipidemia [42], age [41, 44–46], male [47] and traditional risk factors that predict an increased risk of stroke and coronary artery disease in SLE [48–50]. Libman-Sacks endocarditis is one of the complications in SLE and could be a significant risk for embolic cerebrovascular disease [51].

The definition of cerebrovascular disease in the ACR nomenclature is "neurologic deficits due to arterial insufficiency or occlusion, venous occlusive disease, or hemorrhage". These are mainly focal deficits but may be multifocal in recurrent disease. *Diagnostic criteria* is defined as satisfying one of the following and supporting radioimaging study: (1) Stroke syndrome, (2) Transient ischemic attack, (3) Chronic multifocal disease, (4) Subarachnoid and intracranial hemorrhage, and (5) Sinus thrombosis. The finding of unidentified bright objects on MRI without clinical manifestations is not classified for cerebrovascular disease at the present time. In order to confirm the diagnosis of cerebrovascular disease, neuroimaging study, especially MRI, is extremely useful. However, it should be remembered that reversible focal neurological deficit, presumably due to vasculitis [52], might be misdiagnosed as cerebrovascular disease that is irreversible.

5.3.2.3 Demyelinating Syndrome.

Demyelinating syndrome must have the evidence of discrete neurologic lesions distributed in place and time like multiple sclerosis (MS) [53] and is very rare complication seen in about 0.3% of NPSLE patients [2, 54]. Recently, neuromyelitis optica spectrum disorder (NMOSD) has been recognized as a clinical entity independent from MS through the discovery of specific autoantibodies [55]. Thus, NMOSD has been regarded as demyelinating syndrome. In fact, our patient who had showed remission and relapse of upper cervical spinal cord lesions was diagnosed as demyelinating syndrome, has recently turned out to be positive for serum anti-Aquaporin 4 antibody. Thus, this patient should have been diagnosed as NMOSD (Fig. 5.2).

The definition of demyelinating syndrome in the ACR nomenclature is "acute or relapsing demyelinating encephalomyelitis with evidence of discrete neurologic lesions distributed in place and time". *Diagnostic criteria* demand for the presence of neurological lesions occurring at multiple occasions in place and time. It should be noted that transverse myelopathy, optic neuropathy and other cranial nerve

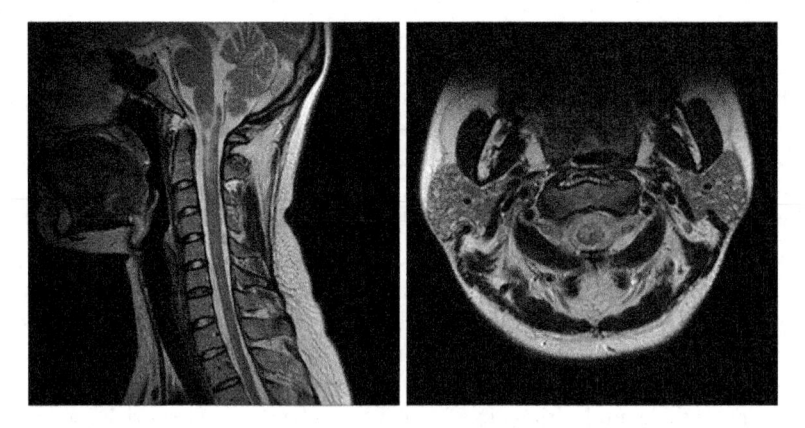

Fig. 5.2 An SLE patient with NMOSD-like lesions. A 20-year-old female with positive anti-serum Aquaporin 4 antibody presenting extensive cervical lesion on MRI

palsies are included in the diagnostic criteria. Since these are also listed as a separate entity in the ACR nomenclature and case definitions, it is somewhat confusing. Although ACR recommends that patients who meet criteria for these and for demyelinating syndrome should be classified as having both, revision with more sophisticated classification would be necessary.

Although spinal cord lesions attract more attention, there is definitely a case which is most appropriately classified as demyelinating syndrome, presenting diffuse hyperintensity in the cerebral white matter along with patchy nodules (Fig. 5.3).

5.3.2.4 Headache

Headache is very common in patients with SLE (32–75.7%) [56–58], although it is also common in general population. Among various types of headache, 38% of patients had migraine and 36% of had tension-type headache [58]. It remains unclear which type of headache is caused by immunological processes related with SLE and requires immunosuppressive therapy. Of note, it has been found that headache due to intracranial hypertension and intractable non-specific headache, but not migraine, are characterized by the inflammatory profile in CSF, such as the elevation of IL-6, IL-8, and IP-10 [59].

The definition of headache in the ACR nomenclature is "discomfort in the region of the cranial vault". Headache in NPSLE can be classified as follows: (1) Migraine with or without aura, (2) Tension headache (episodic tension type headache), (3) Cluster headache, (4) Headache from intracranial hypertension (Pseudotumor cerebri, benign intracranial hypertension) and (5) Intractable headache, nonspecific. It should be noted that intractable headache might be associated with anti-phospholipid antibodies.

FLAIR images on MRI

Remission   **Exacerbation**

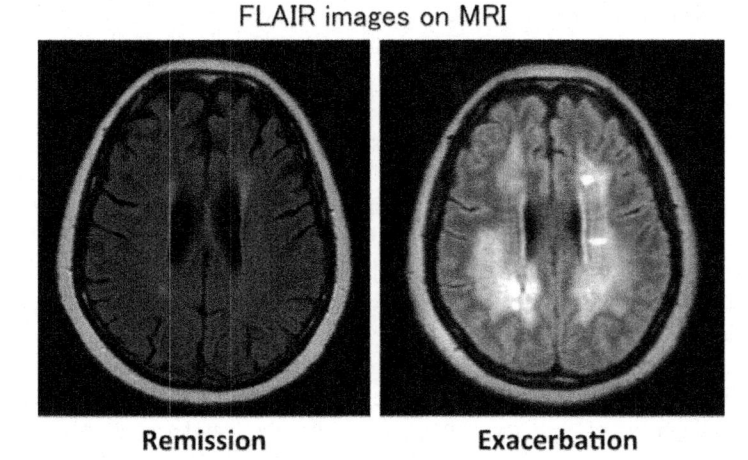

Fig. 5.3 An SLE patient with demyelinating syndrome. A 32-year-old female presented gait disturbance and bladder-bowel disturbances. Compared with the remission phase (left), diffuse hyperintensity in cerebral white matter with scattered nodular lesions are observed in FLAIR images on MRI (right)

5.3.2.5 Movement Disorder (Chorea)

Chorea is the most common among movement disorders [60], affecting bilaterally and symmetrically, whereas hemiballism and Parkinsonian disorders are rare [61, 62]. Movement disorder is usually accompanied by anti-phospholipid antibodies [61–63].

The definition of chorea is "irregular, involuntary and jerky movements, that may involve any portion of the body in random sequence". Each movement is brief and unpredictable. *Diagnostic criteria* require both of the following: A. Observed abnormal movements, B. Random, unpredictable sequence of movements. MRI of the brain is sometimes useful to identify the responsible lesion.

5.3.2.6 Myelopathy

In the 1999 ACR nomenclature and case definitions, myelopathy is *defined* as "disorder of the spinal cord characterized by rapidly evolving paraparesis and/or sensory loss, with a demonstrable motor and/or sensory cord level (may be transverse) and/or sphincter involvement " [3]. *Diagnostic criteria* include usually rapid onset (hours or days) of one or more of the following: A. Bilateral weakness of legs with or without arms (paraplegia/quadriplegia); may be asymmetric. B. Sensory impairment with cord level similar to that of motor weakness; with or without bowel and bladder dysfunction.

As mentioned in the section of demyelinating syndrome, myelopathy is somewhat an obscure word, which might imply various conditions of different etiology, including transverse myelitis due to vasculitis, NMOSD and ischemic myelopathy due to anti-phospholipid syndrome. Therefore, a more sophisticated classification would be required as is also in the case with demyelinating syndrome.

5.3.2.7 Seizures and Seizure Disorders

Seizure is seen in about 10% of SLE patients [64, 65]. Epileptic seizure is common, whereas most seizures are isolated [64]. Of note, isolated seizures might complicate in patients with diffuse NPSLE, especially ACS. Seizures mainly develop during early phase of SLE [66, 67], and are found to be associated with anti-phospholipid antibodies [68].

The definition in the ACR nomenclature is "abnormal paroxysmal neuronal discharge in the brain causing abnormal function". Isolated seizures are distinguished from the diagnosis of epilepsy. Epilepsy is a chronic disorder characterized by an abnormal tendency for recurrent, unprovoked seizures that are usually stereotypic. *Diagnostic criteria* include the following: A. Independent description by a reliable witness, B. Electroencephalogram (EEG) abnormalities. For *diagnosis* of epilepsy, EEG is very useful due to its sensitive, but must be used with clinical data. In most patients with epilepsy, interictal EEG is normal. Seizures are divided into partial and generalized (Table 5.4). Partial seizures have clinical or electroencephalographic evidence of a focal onset; the abnormal discharge usually arises in a portion of one hemisphere and may spread to the rest of the brain. Primary generalized seizures have no interictal evidence on EEG of focal onset. A generalized seizure can be primary or secondary.

5.3.3 Neurologic Syndromes of PNS

Neurologic syndromes of the PNS in SLE (PNS-NPSLE) is disturbance of the cranial and peripheral nerves. Peripheral neuropathy attributed to SLE occurs in 1.5–17.7% [69–72]. Importantly, peripheral neuropathy can be caused by many

Table 5.4 Clinical types of seizure disorders

Seizures and seizure disorders
Primary generalized seizure (bilaterally symmetric and without local onset)
• Tonic clonic (grand mal) or tonic or clonic
• Atonic or astatic seizure
• Absence seizure (petit mal)
• Myoclonic seizures
Partial or focal seizures (seizures begining locally) (also referred to as Jacksonian, temporal lobe or psychomotor seizure, according to type)
• Simple without impairment of consciousness
• Complex with partial impairment of consciousness
• Simple or complex

5 Clinical Features 71

conditions other than SLE. In addition, symptoms like myasthenia and numbness, mostly common in PNS-NPSLE, are non-specific. Therefore, neuropathy without nerve conduction study and electromyogram (EMG) should not be classified into PNS-NPSLE.

5.3.3.1 Neuropathy, Cranial

As cranial neuropathy, oculomotor palsy has been well documented [73, 74], but should be distinguished from a disease due to brainstem lesions [75]. Optic neuropathy can be developed with transverse myelitis as well at the same time [76].

5.3.3.2 Acute Inflammatory Demyelinating Polyradiculoneuropathy (Guillain-Barré Syndrome)

The definition of acute inflammatory demyelinating polyradiculoneuropathy is "acute, inflammatory, and demyelinating syndrome of spinal roots, peripheral, and occasionally cranial nerves". According to a single center cohort study, PNS-NPSLE was observed in 17.7% of SLE patients, whereas Guillain-Barré Syndrome was seen only in 1.1% of patients with PNS-NPSLE [72].

5.3.3.3 Autonomic Disorder

Autonomic disorder is relatively common in patients with SLE [77–79]. Thus, 18% of the patients with SLE had significantly more abnormal results of autonomic tests by noninvasive autonomic tests, compared with 3% of the controls in prospective study [80]. *The definition* of autonomic disorder is "disorder of the autonomic nervous system with orthostatic hypotension, sphincteric erectile/ejaculatory dysfunction, anhidrosis, heat intolerance, constipation". For diagnosis abnormal response to provocative tests needs to be confirmed.

5.3.3.4 Mononeuropathy (Single/Multiplex)

In a retrospective study, non-compression mononeuropathy was shown in 23.7% of patients with peripheral neuropathy [72]. *The definition* of mononeuropathy in the ACR nomenclature is "disturbed function of one or more peripheral nerve(s) resulting in weakness/paralysis or sensory dysfunction due to either conduction block in motor nerve fibers or axonal loss". Conduction block is due to demyelination with preserved axon continuity. Remyelination may take place rapidly and completely. Once axonal damages take place, axonal degeneration is induced below the damaged site and the recovery is usually slow and incomplete. All modalities or certain forms of sensation may be affected. *Diagnostic criteria* include A. Clinical demonstration of motor/sensory disturbances in the distribution of a peripheral nerve and/or B. Abnormalities on nerve conduction studies or EMG (i.e., concentric needle examination).

5.3.3.5 Myasthenia Gravis

Myasthenia gravis (MG) was found in 7.5% of PNS-NPSLE patients [72]. Among 73 PNS-SLE patients, 3 patients manifested myasthenia gravis before the onset of SLE [71]. However, it has been unclear whether SLE can be a risk of for development of MG. Also, the severity of MG may be associated with concomitant SLE [81]. *The definition* of MG in the ACR nomenclature is "neuromuscular transmission disorder characterized by fluctuating weakness and fatigability of bulbar and other voluntary muscles without loss of reflexes or impairment of sensation or other neurologic function". MG is caused by antibodies against acetylcholine receptors, and therefore could co-exist with other autoimmune diseases such as SLE.

5.3.3.6 Plexopathy

The definition is "disorder of brachial or lumbosacral plexus producing muscle weakness, sensory deficit, and/or reflex change not corresponding to the territory of single root or nerve". Plexopathy is considered to be extremely rare.

5.3.3.7 Polyneuropathy

The definition of polyneuropathy is "acute or chronic disorder of sensory and motor peripheral nerves with variable tempo characterized by symmetry of symptoms and physical findings in a distal distribution". Resent study demonstrated that small-fiber neuropathy is as frequent as peripheral neuropathy in NPSLE [70]. Thus, it was disclosed that 17.1% of the patients with SLE-related peripheral neuropathies had biopsy-proven small-fiber neuropathies [70].

5.4 Summary

The 1999 ACR nomenclature and case definitions of NPSLE has made a milestone in our understanding of NPSLE. However, even this ACR nomenclature has not been covering all the manifestations clinically observed in real lupus patients. As highlighted in this chapter, 19 syndromes of CNS and PNS disorders are defined along with diagnostic criteria for each. Among the manifestations, headache, mood disorders, anxiety and mild cognitive dysfunction can be often found without lupus activity and therefore the evaluation of their attribution and activity is important.

There is an apparent confusion in the classification of demyelinating syndrome and myelopathy, partially because of the discovery of anti-Aquaporin 4 antibody and establishment of the clinical entity of NMOSD. In addition, new categories of manifestations are described, such as reversible focal neurological deficits and small-fiber neuropathy. Further studies for reclassification are absolutely necessary.

Reference

1. Hebra F, et al. On diseases of the skin, including the exanthemata. London: The New Sydenham Society; 1874. p. 14–47.
2. Unterman A, et al. Neuropsychiatric syndromes in systemiclupuserythematosus: a meta-analysis. Semin Arthritis Rheum. 2011;41:1–11.
3. ACR Ad Hoc Committee on Neuropsychiatric Lupus Nomenclature. The American College of Rheumatology nomenclature and case definitions for neuropsychiatric lupus syndromes. Arthritis Rheum. 1999;42:599–608.
4. Zirkzee E, et al. Mortality in neuropsychiatric systemic lupus erythematosus (NPSLE). Lupus. 2014;23:31–8.
5. Tay S, et al. Active disease is independently associated with more severe anxiety rather than depressive symptoms in patients with systemic lupus erythematosus. Lupus. 2015;24:1392–9.
6. Maneeton B, et al. Prevalence and predictors of depression in patients with systemic lupus erythematosus: a cross-sectional study. Neuropsychiatr Dis Treat. 2013;9:799–804.
7. Bertsias GK, et al. EULAR recommendations for the management of systemic lupus erythematosus with neuropsychiatric manifestations: report of a task force of the EULAR standing committee for clinical affairs. Ann Rheum Dis. 2010;69:2074–82.
8. Ainiala H, et al. The prevalence of neuropsychiatric syndromes in systemic lupus erythematosus. Neurology. 2001;57:496–500.
9. Sung SC, et al. The impact of chronic depression on acute and long-term outcomes in a randomized trial comparing selective serotonin reuptake inhibitor Monotherapy versus each of 2 different antidepressant medication combinations. J Clin Psychiatry. 2012;73:967–76.
10. Meszaros ZS, et al. Psychiatric symptoms in systemic lupus erythematosus: a systematic review. J Clin Psychiatry. 2012;73:993–1001.
11. Hanly JG, et al. Cognitive function in systemic lupus erythematosus: results of a 5-year prospective study. Arthritis Rheum. 1997;40:1542–3.
12. Waterloo K, et al. Neuropsychological function in systemic lupus erythematosus: a five-year longitudinal study. Rheumatology (Oxford). 2002;41:411–5.
13. Hanly JG, et al. Management of neuropsychiatric lupus. Best Pract Res Clin Rheumatol. 2005;19:799–821.
14. Brey RL, et al. Neuropsychiatric syndromes in lupus: prevalence using standardized definitions. Neurology. 2002;58:1214–20.
15. Sanna G, et al. Neuropsychiatric manifestations in systemic lupus erythematosus: prevalence and association with antiphospholipid antibodies. J Rheumatol. 2003;30:985–92.
16. Hanly JG, et al. Neuropsychiatric events in systemic lupus erythematosus: attribution and clinical significance. J Rheumatol. 2004;31:2156–62.
17. Mok CC, et al. Neuropsychiatric manifestations and their clinical associations in southern Chinese patients with systemic lupus erythematosus. J Rheumatol. 2001;28:766–71.
18. Tomietto P, et al. General and specific factors associated with severity of cognitive impairment in systemic lupus erythematosus. Arthritis Rheum. 2007;57:1461–72.
19. Lim L, et al. Psychiatric and neurological manifestations in systemic lupus erythematosus. Q J Med. 1988;66:27–38.

20. Harboe E, et al. Fatigue is associated with cerebral white matter hyperintensities in patients with systemic lupus erythematosus. J Neurol Neurosurg Psychiatry. 2008;79:199–201.
21. Kozora E, et al. Major life stress, coping styles, and social support in relation to psychological distress in patients with systemic lupus erythematosus. Lupus. 2005;14:363–72.
22. Wekking EM. Psychiatric symptoms in systemic lupus erythematosus: an update. Psychosom Med. 1993;55:219–28.
23. Bachen EA, et al. Prevalence of mood and anxiety disorders in women with systemic lupus erythematosus. Arthritis Rheum. 2009;61:822–9.
24. Líndal E, et al. Psychiatric disorders among subjects with systemic lupus erythematosus in an unselected population. Scand J Rheumatol. 1995;24:346–51.
25. Pego-Reigosa JM, et al. Psychosis due to systemic lupus erythematosus: characteristics and long-term outcome of this rare manifestation of the disease. Rheumatology (Oxford). 2008;47:1498–502.
26. Kohen M, et al. Lupus psychosis: differentiation from the steroid-induced state. Clin Exp Rheumatol. 1993;11:323–6.
27. Denburg SD, et al. Corticosteroids and neuropsychological functioning in patients with systemic lupus erythematosus. Arthritis Rheum. 1994;37:1311–20.
28. Wysenbeek AJ, et al. Acute central nervous system complications after pulse steroid therapy in patients with systemic lupus erythematosus. J Rheumatol. 1990;17:1695–6.
29. Chau SY, et al. Factors predictive of corticosteroid psychosis in patients with systemic lupus erythematosus. Neurology. 2003;61:104–7.
30. Shimizu Y, et al. Post-steroid neuropsychiatric manifestations are significantly more frequent in SLE compared with other systemic autoimmune diseases and predict better prognosis compared with de novo neuropsychiatric SLE. Autoimmun Rev. 2016;15:786–94.
31. Sergent JS, et al. Central nervous system disease in systemic lupus erythematosus. Therapy and prognosis. Am J Med. 1975;58:644–54.
32. Baizabal-Carvallo JF, et al. Clinical characteristics and outcomes of the meningitides in systemic lupus erythematosus. Eur Neurol. 2009;61:143–8.
33. Sands ML, et al. Recurrent aseptic meningitis followed by transverse myelitis as a presentation of systemic lupus erythematosus. J Rheumatol. 1988;15:862–4.
34. Suzuki Y, et al. Severe cerebral and systemic necrotizing vasculitis developing during pregnancy in a case of systemic lupus erythematosus. J Rheumatol. 1990;17:1408–11.
35. Marin OS, et al. Cerebral vasculitis. Presenting as a meningoencephalitis. A case of systemic lupus erythematosus. Del Med J. 1973;45:315–9.
36. Ostensen M, et al. Nonsteroidal anti-inflammatory drugs in systemic lupus erythematosus. Lupus. 2000;9:566–72.
37. Escalante A, et al. Trimethoprim-sulfamethoxasole induced meningitis in systemic lupus erythematosus. J Rheumatol. 1992;19:800–2.
38. Bruner KE, et al. Trimethoprim-sulfamethoxazole-induced aseptic meningitis-not just another sulfa allergy. Ann Allergy Asthma Immunol. 2014;113:520–6.
39. Sciascia S, et al. The estimated frequency of antiphospholipid antibodies in young adults with cerebrovascular events: a systematic review. Ann Rheum Dis. 2015;74:2028–33.
40. Petri M, et al. Derivation and validation of the systemic lupus international collaborating clinics classification criteria for systemic lupus erythematosus. Arthritis Rheum. 2012;64:2677–86.
41. Holmqvist M, et al. Stroke in systemic lupus erythematosus: a meta-analysis of population-based cohort studies. RMD Open. 2015;1:e000168.
42. Kitagawa Y, et al. Stroke in systemic lupus erythematosus. Stroke. 1990;21:1533–9.
43. Mikdashi J, et al. Baseline disease activity, hyperlipidemia, and hypertension are predictive factors for ischemic stroke and stroke severity in systemic lupus erythematosus. Stroke. 2007;38:281–5.
44. Toloza SMA, et al. Systemic lupus erythematosus in a multiethnic US cohort (LUMINA). XXIII. Baseline predictors of vascular events. Arthritis Rheum. 2004;50:3947–57.
45. Csépány T, et al. MRI findings in central nervous system systemic lupus erythematosus are associated with immunoserological parameters and hypertension. J Neurol. 2003;250:1348–54.

46. Morelli S, et al. Left-sided heart valve abnormalities and risk of ischemic cerebrovascular accidents in patients with systemic lupus erythematosus. Lupus. 2003;12:805–12.
47. Petri M, et al. Plasma homocysteine as a risk factor for atherothrombotic events in systemic lupus erythematosus. Lancet. 1996;348:1120–4.
48. Esdaile JM, et al. Traditional Framingham risk factors fail to fully account for accelerated atherosclerosis in systemic lupus erythematosus. Arthritis Rheum. 2001;44:2331–7.
49. Bessant R, et al. Risk of coronary heart disease and stroke in a large British cohort of patients with systemic lupus erythematosus. Rheumatology (Oxford). 2004;43:924–9.
50. Urowitz MB, et al. Atherosclerotic vascular events in a multinational inception cohort of systemic lupus erythematosus. Arthritis Care Res (Hoboken). 2010;62:881–7.
51. Roldan CA, et al. Libman-sacks endocarditis and embolic cerebrovascular disease. JACC Cardiovasc Imaging. 2013;6:973–83.
52. Kimura M, et al. Reversible focal neurological deficits in systemic lupus erythematosus: report of 2 cases and review of the literature. J Neurol Sci. 2008;272:71–6.
53. Magro Checa C, et al. Demyelinating disease in SLE: is it multiple sclerosis or lupus? Best Pract Res Clin Rheumatol. 2013;27:405–24.
54. Bertsias GK, et al. Pathogenesis, diagnosis and management of neuropsychiatric SLE manifestations. Nat Rev Rheumatol. 2010;6:358–67.
55. Wingerchuk DM, et al. International consensus diagnostic criteria for neuromyelitis optica spectrum disorders. Neurology. 2015;85:177–89.
56. Hanly JG, et al. Headache in systemic lupus erythematosus: results from a prospective, international inception cohort study. Arthritis Rheum. 2013;65:2887–97.
57. Mitsikostas DD, et al. A meta-analysis for headache in systemic lupus erythematosus: the evidence and the myth. Brain. 2004;127:1200–9.
58. Omdal R, et al. Somatic and psychological features of headache in systemic lupus erythematosus. J Rheumatol. 2001;28:772–9.
59. Fragoso-Loyo H, et al. Inflammatory profile in cerebrospinal fluid of patients with headache as a manifestation of neuropsychiatric systemic lupus erythematosus. Rheumatology (Oxford). 2013;52:2218–22.
60. Baizabal-Carvallo JF, et al. Movement disorders in systemic lupus erythematosus and the antiphospholipid syndrome. J Neural Transm. 2013;120:1579–89.
61. Cervera R, et al. Chorea in the antiphospholipid syndrome. Clinical, radiologic, and immunologic characteristics of 50 patients from our clinics and the recent literature. Medicine (Baltimore). 1997;76:203–12.
62. Asherson RA, et al. Antiphospholipid antibodies and chorea. J Rheumatol. 1988;15:377–9.
63. Orzechowski NM, et al. Antiphospholipid antibody-associated chorea. J Rheumatol. 2008;35:2165–70.
64. González-Duarte A, et al. Clinical description of seizures in patients with systemic lupus erythematosus. Eur Neurol. 2008;59:320–3.
65. Appenzeller S, et al. Epileptic seizures in systemic lupus erythematosus. Neurology. 2004;63:1808–12.
66. Andrade RM, et al. Seizures in patients with systemic lupus erythematosus: data from LUMINA, a multiethnic cohort (LUMINA LIV). Ann Rheum Dis. 2008;67:829–34.
67. Herranz MT, et al. Association between antiphospholipid antibodies and epilepsy in patients with systemic lupus erythematosus. Arthritis Rheum. 1994;37:568–71.
68. Mackworth-Young CG, et al. Epilepsy: an early symptom of systemic lupus erythematosus. J Neurol Neurosurg Psychiatry. 1985;48:185.
69. Florica B, et al. Peripheral neuropathy in patients with systemic lupus Erythematosus. Semin Arthritis Rheum. 2011;41:203–11.
70. Oomatia A, et al. Peripheral neuropathies in systemic lupus erythematosus: clinical features, disease associations, and immunologic characteristics evaluated over a twenty-five-year study period. Arthritis Rheumatol. 2014;66:1000–9.
71. Xianbin W, et al. Peripheral neuropathies due to systemic lupus erythematosus in China. Medicine (Baltimore). 2015;94:e625.

72. Toledano P, et al. Peripheral nervous system involvement in systemic lupus erythematosus: prevalence, clinical and immunological characteristics, treatment and outcome of a large cohort from a single centre. Autoimmun Rev. 2017;16:750–5.
73. Keane JR. Eye movement abnormalities in systemic lupus erythematosus. Arch Neurol. 1995;52:1145–9.
74. Saleh Z, et al. Cranial nerve VI palsy as a rare initial presentation of systemic lupus erythematosus: case report and review of the literature. Lupus. 2010;19:201–5.
75. Sivaraj RR, et al. Ocular manifestations of systemic lupus erythematosus. Rheumatology (Oxford). 2007;46:1757–62.
76. Feinglass EJ, et al. Neuropsychiatric manifestations of systemic lupus erythematosus: diagnosis, clinical spectrum, and relationship to other features of the disease. Medicine (Baltimore). 1976;55:323–39.
77. Omdal R, et al. Autonomic function in systemic lupus erythematosus. Lupus. 1994;3:413–7.
78. Gamez-Nava JI, et al. Autonomic dysfunction in patients with systemic lupus erythematosus. J Rheumatol. 1998;25:1092–6.
79. Hogarth MB, et al. Cardiovascular autonomic function in systemic lupus erythematosus. Lupus. 2002;11:308–12.
80. Shalimar, et al. Autonomic dysfunction in systemic lupus erythematosus. Rheumatol Int. 2006;26:837–40.
81. Bekircan-Kurt CE, et al. The course of myasthenia gravis with systemic lupus erythematosus. Eur Neurol. 2014;72:326–9.

Chapter 6
Cytokines and Chemokines

Taku Yoshio and Hiroshi Okamoto

Abstract Neuropsychiatric syndromes of systemic lupus erythematosus (NPSLE) is a life-threatening disorder and early diagnosis and proper treatment are critical for the management of patients with this disease. Brain magnetic resonance imaging, electroencephalogram, neuropsychological tests and routine cerebrospinal fluid (CSF) examination are used clinically for the diagnosis of NPSLE. In addition to these tests, cytokine and chemokine levels in the CSF have been reported as useful diagnostic markers of NPSLE. This chapter provides an overview of the roles of cytokines and chemokines in NPSLE.

Keywords NPSLE · Cytokines · Chemokines · BBB · CSF · IL-6 · IL-8 · MCP-1 · IP-10 · G-CSF · TNF-α · IL-10 · IFN-α · RANTES · Fractalkine · The IP-10/MCP-1 ratio

6.1 Introduction

Systemic lupus erythematosus (SLE) is an autoimmune disease characterized by widespread immunologic abnormalities and multi-organ involvement, including the skin, joints, and kidney, as well as the peripheral and central nervous systems (CNS). Neuropsychiatric syndromes in systemic lupus erythematosus (NPSLE) may occur at any time during the course of the disease, and symptoms are extremely diverse, ranging from depression, psychosis, and seizures to stroke [1]. The origin of minor clinical symptoms, such as headaches and mood swings are not specific for NPSLE. In fact, SLE patients may be under the influence of other conditions

T. Yoshio (✉)
Division of Rheumatology and Clinical Immunology, School of Medicine, Jichi Medical University, Shimotsuke-shi, Tochigi, Japan
e-mail: takuyosh@jichi.ac.jp

H. Okamoto
Minami-otsuka institute of technology, Minami-otsuka Clinic, Tokyo, Japan

© Springer International Publishing AG, part of Springer Nature 2018
S. Hirohata (ed.), *Neuropsychiatric Systemic Lupus Erythematosus*,
https://doi.org/10.1007/978-3-319-76496-2_6

capable of causing neuropsychiatric symptoms, such as infections, severe hypertensions, metabolic complications, steroid psychosis, and other drug toxicities [2]. Without proper treatment, neuropsychiatric involvement in SLE is known to increase morbidity and mortality, and therefore the availability of beneficial treatments increases the need for the early recognition of neuropsychiatric manifestations in SLE. Along with more specific diagnostic tools and an effective method of monitoring disease activity, therapeutic responses are crucial in the management of NPSLE. Currently, tests for diagnosing NPSLE include brain magnetic resonance imaging, electroencephalogram, neuropsychological tests and routine cerebrospinal fluid (CSF) examination. The results of these tests are reported to be abnormal in some but not all patients, and therefore none of the findings are specific for NPSLE. The large discrepancy in the reported frequency of neuropsychiatric involvement in SLE patients (14–75%) further proves there is no single confirmatory diagnostic tool [3, 4].

Increased levels of proinflammatory cytokines and chemokines have been reported in the CSF of patients with NPSLE. Thus, several reports have shown cytokines and chemokines, such as interleukin (IL)-6, IL-1, IL-8/CXCL8, IL-10, tumor necrosis factor (TNF)-α, interferon (IFN)- α, IFN- γ, monocyte chemotactic protein 1 (MCP-1)/CCL2, interferon-gamma inducible protein-10 (IP-10)/CXCL10, Fractalkine/CX3CL1 and granurocyte-colony stimulating factor (G-CSF), to be elevated intrathecally, thereby allowing these cytokines and chemokines to be used as diagnostic tools [5–10].

Cytokines and chemokines are considered to be therapeutic targets in several chronic inflammatory disorders such as SLE. Based on the number of recently published studies, this chapter focuses on the use of cytokines and chemokines as biomarkers as well as pathogenic factors in NPSLE.

6.2 The Blood-Brain Barrier

The blood-brain barrier (BBB) is a highly specialized, multi-cellular structure that functions as a selective diffusion barrier between the peripheral circulation and the CNS. The BBB is composed of specialized endothelial cells (ECs) that are linked by complex tight junctions and adherens junctions. These cells are also surrounded by astrocytes and pericytes. Under normal conditions, the specialized structure of the BBB hinders paracellular transport of most hydrophilic compounds across the cerebral endothelium and restricts migration of blood-borne cells into the CNS. As a result, microglia, the resident immune cells of the CNS, are the initial responders to pathogens or tissue damage. However, prolonged tissue insult triggers inflammatory conditions that cause the BBB to lose its restrictive features, resulting in the subsequent infiltration of peripheral immune cells.

Reactive microglia, astrocytes, and pericytes, as well as ECs, release numerous molecules that promote invasion of peripheral immune cells into the CNS. Secreted inflammatory mediators, including IL-8/CXCL8, MCP-1/CCL2, TNF-α, IL-1β, recruit immune cells and stimulate the expression of adhesion molecules on ECs that participate in integrin-mediated leukocyte tethering, rolling and activation. These pro-inflammatory molecules also trigger the dynamic reorganization of junction complexes between ECs, thereby promoting the formation of paracellular gaps. Matrix metalloproteases, which are also released, degrade proteins present in the extracellular matrix and may contribute to the loss of pericytes. These events lead to an increase in the permeability of the BBB and invasion of peripheral immune cells.

6.3 Cytokines

Cytokines are small substances secreted by specific cells of the immune system which mediate local communication between cells and play important roles in the development and functioning of both the innate and adaptive immune response.

Several cytokines such as IL-1, soluble IL-2R, IL-6, IL-10, TNF-α, IFN-γ, IFN-α, and G-CSF have been reported to be elevated in the CSF from patients with NPSLE [5, 10–22]. A summary of the reported results (IL-1, soluble IL-2R, IL-6, IL-10, TNF-α, IFN-γ, IFN-α, and G-CSF) is shown in Table 6.1.

6.4 Cytokines as Biomarkers

In this section, the role and diagnostic tools of respective cytokine as a biomarker in NPSLE are described.

6.4.1 Tumor Necrosis Factor

The role of TNF-α in lupus is still controversial. TNF-α may be protective in patients with lupus, since low TNF-α activity is associated with increased disease activity. Some patients with rheumatoid arthritis who were treated with anti-TNF-α antibodies, expressed anti-double-stranded DNA antibodies, and even lupus developed in a few of these patients. By contrast, TNF-α may promote the pathogenesis of lupus, since the level of TNF-α messenger RNA was high in kidney-biopsy specimens from patients with lupus nephritis and there is a report showing that giving the anti-TNF-α antibody agent, infliximab, to six patients with lupus led to resolution of joint swelling in three patients with arthritis and a 60% reduction of urinary protein loss in four patients with renal lupus [22, 23].

There is a report studying the expression of IL-4, IL-10, TNF-α and IFN-γ in both peripheral blood lymphocytes (PBLs) and CSF from NPSLE patients whereby the authors found that mRNA for IL-10, TNF-α and IFN-γ were increased in PBLs while only IL-10 and IFN-γ were elevated in CSF [6]. Our group showed that the

Table 6.1 Cytokines in CSF of NPSLE

Cytokines	NPSLE	Control group	Authors	Year
IL-1	Increased	None	Alcocer-Varela et al. [11]	1992
IL-1β	Same	Neulogical symptoms without neurological diseases	Gilad et al. [12]	1997
Soluble IL-2R	Increased	Neulogical symptoms without neurological diseases	Gilad et al. [12]	1997
IL-6	Increased	Non-NPSLE, HI	Jonsen et al. [13]	2003
	Increased	HI, neurocysticerosis	Jara et al. [14]	1998
	Increased	Non-NPSLE	Hirohata and Miyamoto [15]	1990
	Increased	Cerebral infarction	Hirohata and Hayakawa [16]	1999
	Increased	None	Alcocer-Varela et al. [11]	1992
	Increased	CNS inflammation, non-inflammatory CNS diseases	Tsai et al. [17]	1994
	Increased	Non-NPSLE	Trysberg et al. [5]	2009
	Increased	HI	Dellalibera-Joviliano et al. [18]	2003
	Increased	Non-NPSLE, SM, non-AID	Fragoso-Loyo et al. [19]	2007
	Increased	Non-NPSLE	Yoshio et al. [10]	2016
IL-10	Same	Non-NPSLE, HI	Jonsen et al. [13]	2003
	Increased	HI	Dellalibera-Joviliano et al. [18]	2003
	Same	Non-NPSLE, SM, non-AID	Fragoso-Loyo et al. [19]	2007
TNF-α	Increased	HI	Dellalibera-Joviliano et al. [18]	2003
	Same	Neulogical symptoms without neurological diseases	Gilad et al. [12]	1997
	Same	Non-NPSLE, SM, non-AID	Fragoso-Loyo et al [19]	2007
IFN-γ	Same	Non-NPSLE, SM, non-AID	Fragoso-Loyo et al. [19]	2007
IFN-α	Increased	Non-NPSLE, HI	Jonsen et al. [13]	2003
	Increased	Non-NPSLE	Winfield et al. [20]	1983
	Increased	Non-NPSLE	Shiozawa et al. [21]	1992

AID autoimmune diseases, *HI* healthy individuals, *MS* multiple sclerosis, *SM* septic meningitis

mean CSF levels of TNF-α were significantly higher in the 30 patients with central NPSLE as compared to the 22 non-NPSLE patients. However, a comparison of cytokine and chemokine levels in the CSF and serum samples of 30 patients with central NPSLE from whom the CSF and serum samples were obtained at the same time showed that CSF TNF-α levels were much lower than serum TNF-α levels [10]. One report showed that CSF TNF-α levels in patients with NPSLE were higher than those in healthy controls [18], however two other reports did not show an increase of CSF TNF-α levels in patients with NPSLE [12, 19]. It is therefore uncertain whether CSF TNF-α is associated with the pathogenesis of NPSLE.

6.4.2 Interleukin-10

Serum levels of IL-10 are consistently high in patients with lupus, and they correlate with the activity of the disease. IL-10 has a number of biologic effects, including stimulation of polyclonal populations of B lymphocytes. In fact, blocking this cytokine could reduce the production of pathogenic autoantibodies [23].

There is a report studying the expression of IL-4, IL-10, TNF-α and IFN-γ in both PBLs and CSF from NPSLE patients whereby the authors found that mRNA levels for IL-10, TNF-α and IFN-γ were increased in PBLs while only IL-10 and IFN-γ were elevated in CSF [6]. Our group showed that the mean CSF levels of IL-10 were significantly higher in the 30 patients with central NPSLE as compared to those in the 22 non-NPSLE patients. Interestingly, these results are in contrast to a previous study which demonstrated that CSF IL-10 levels were significantly lower than serum IL-10 levels [10]. One report showed that CSF IL-10 levels in patients with NPSLE were higher than those in healthy controls [18], however two other reports did not show increased CSF IL-10 levels in patients with NPSLE [13, 19]. Again, it is thus controversial whether CSF IL-10 is associated with the pathogenesis of NPSLE.

6.4.3 Interferon-α

Serum levels of interferon-α (IFN-α) are elevated in patients with active lupus and microarray studies showed that 13 genes regulated by IFN were up-regulated in peripheral-blood mononuclear cells from patients with lupus, as compared with healthy controls [23].

IFN-α was also detected in the CSF of patients with NPSLE, and is of particular interest in the pathophysiology of NPSLE, given its ability to promote an autoimmune response and its recognized role in the etiopathogenesis of SLE [13, 20, 21].

As SLE is an autoimmune disorder characterized by numerous autoantibodies, a pathogenetic role for autoantibodies is theoretically suspected. Immune complexes in SLE can stimulate IFN-α production and there is strong evidence in humans and in mice that IFN- α can cause neuropsychiatric manifestations. Santer DM et al. used a bioassay containing plasmacytoid dendritic cells to demonstrate that NPSLE CSF induced significantly higher IFN-α compared with CSF from patients with multiple sclerosis or other autoimmune disease controls [24]. When normalized for IgG concentration, NPSLE CSF was 800-fold more potent at inducing IFN- α compared with paired serum, due to inhibitors present in serum. In addition to IFN-α, immune complexes formed by CSF autoantibodies produced significantly increased levels of IP-10/CXCL, IL-8/CXCL8 and MCP-1/CCL2. From these results they proposed a two-step model of NPSLE whereby CSF autoantibodies bind to antigens released by neurocytotoxic antibodies or other brain cell injury, and the resulting immune complexes stimulate IFN-α, proinflammatory cytokines and chemokines [24].

Indirect support for the role of IFN-α in NPSLE comes from the untoward side effects of this cytokine when used as a therapeutic modality for treatment of hepatitis or malignancy, with approximately one third of patients receiving IFN-α exhibiting CNS symptoms [25]. Although depression is the most common feature, other symptoms, such as psychosis, confusion, mania, and seizures have also been reported. Of note, IL-6 may potentiate the depressive propensity of IFN-α, as high serum levels of IL-6 prior to administration of IFN-α has been reported to predict the development of depression [26].

6.4.4 Interleukin-6

Among reported cytokines, IL-6 has been shown to have the strongest positive association with NPSLE. An exhaustive study of cytokines and chemokines recently reported that IL-6 and IL-8/CXCL8 were elevated in NPSLE compared with non-NPSLE and non-autoimmune disease patients [19]. This study also found that IL-2, IL-4, IL-10, TNF-α, and IFN-γ were low in all groups examined [19]. In other reports, no association was found between IL-2, IL-6, IL-10, TNF-α and IFN -γ with NPSLE [12, 13]. A recent study showed that the sensitivity and specificity of CSF IL-6 for diagnosis of lupus psychosis was 87.5% and 92.3%, respectively, indicating that CSF IL-6 might be an effective marker for the diagnosis of lupus psychosis [27].

More recently, a comparison of cytokine and chemokine levels in the CSF and serum samples of 30 patients with central NPSLE from whom the CSF and serum samples were obtained at the same time, suggested that the intrathecal concentrations of IL-6, IL-8/CXCL8, IP-10/CXCL10, MCP-1/CCR2 and G-CSF were not influenced by the serum concentrations in patients with central NPSLE [10]. These data indicated that production of these cytokines and chemokines might take place in the CNS. To confirm the role of these small molecules in the pathogenesis of NPSLE, the levels of IL-6, IL-8/CXCL8, IP-10/CXCL10, MCP-1/CCL2 and G-CSF in the CSF from 30 patients with central NPSLE were compared with those in 22 patients with non-NPSLE. The mean levels of CSF IL-6, IL-8/CXCL8, IP-10/CXCL10, MCP-1/CCL2 and G-CSF were significantly higher in 30 patients with central NPSLE as compared to those in 22 patients with non-NPSLE [10]. Importantly, the largest differences occurred in the level of IL-6 in the CSF [10]. Thus IL-6 in the CSF might be the most useful diagnostic marker of central NPSLE among the cytokines and chemokines investigated.

IL-6 levels in the CSF of NPSLE were reported to be elevated without damage of the BBB. In addition, the expression of IL-6 mRNA was elevated in the hippocampus and cerebral cortex, suggesting that IL-6 expression was increased within the entire CNS of NPSLE [15, 16].

On the other hand, our group demonstrated the *in vitro* activation of human ECs by anti-NR2 glutamate receptor antibodies (anti-NR2) (the enhanced production of cytokines such as IL-6 and IL-8/CXCL8 and the up-regulated expression of adhesion molecules such as ELAM-1, ICAM-1 and VCAM-1) through the activation of the NF-kB pathway [28]. Consistently, the production of IL-6, IL-8, IP-10, MCP-1 and G-CSF by ECs has been reported [28–31]. Therefore, the BBB damage might be caused by autoantibodies such as anti-NR2 or antiribosomal P protein antibodies that react with ECs in NPSLE patients. Thus, this damage in ECs of the BBB leads to the increased concentrations of IL-6, IL-8/CXCL8, IP-10/CXCL10, MCP-1/CCR2 and G-CSF in the CSF, allowing access to the CNS by autoantibodies, immune complexes and immune cells such as leukocytes in the circulation, resulting in inflammation in the brain.

In addition, intrathecal production of these cytokines and chemokines by neuronal or glial cells might also take place. Furthermore, these cytokines and chemokines might increase the permeability of the BBB, thus providing access to the CNS for autoantibodies, immune complexes and immune cells such as leukocytes in the circulation. It is conceivable that both the degree of the BBB dysfunction and the type and titer of autoantibodies might be the determining factors in the development of certain diffuse NPSLE, such as psychosis and acute confusional state.

The TNF family ligands BAFF (B-cell activating factor of TNF family) and APRIL (a proliferation-inducing ligand) are essential for B-cell proliferation, differentiation and function. Intrathecal IL-6 in NPSLE is associated with the CSF immunoglobulin (Ig) G Index, a measurement of intrathecal IgG production, suggesting that IL-6 in concert with BAFF and APRIL, which are also elevated in CSF from patients with diffuse NPSLE, may increase B-cell activation within the CNS [32].

Elevated serum levels of BAFF and APRIL have been reported in patients with SLE. Recently BAFF and APRIL were studied in the CSF of NPSLE patients. They found that levels of APRIL in CSF were more than 20-fold higher and levels of BAFF in CSF were more than 200-fold higher than those of healthy controls [33]. Comparing the levels of APRIL in CSF between NPSLE and non-NPSLE patients, enhanced levels of APRIL were noted in NPSLE. Moreover, they found that CSF levels of APRIL correlated with BAFF but not with IL-6 [33].

There is a report regarding the association between cytokine levels and acute confusional state (ACS) of NPSLE [32]. The authors performed a prospective study using a cohort of 59 patients with SLE and compared patients with and without ACS as well as associations between ACS and each CSF test (IL-6, IL-8/CXCL8, IFN-α, IgG index, and Q-albumin). In this study, ACS was diagnosed in 10 patients (ACS group), NPSLE except ACS in 13 patients, and non-NPSLE in 36 patients (non-NPSLE group). CSF IL-6 levels in the ACS group were significantly higher than those in the non-NPSLE group ($p < 0.05$) and a positive IgG index ($p = 0.028$) was significantly associated with ACS. No other test showed a significant association with ACS. The positive and negative predictive values for the diagnosis of ACS in SLE were 80% and 85% for elevated CSF IL-6 levels (greater than 31.8 pg/ml), and 75% and 83% for the IgG index, respectively. From these results, the authors concluded that no single CSF test had sufficient predictive value to diagnose ACS in SLE, although CSF IL-6 levels and the IgG index showed statistical associations with ACS [32].

Increased levels of intrathecal IL-6 have been reported in numerous inflammatory conditions, such as other autoimmune diseases (Neuro-Behçet's syndrome, mixed connective tissue disease) and neurologic conditions such as CNS infections, cerebrovascular events and myelitis [34, 35]. Therefore, the possibility of these conditions must be excluded to confirm that a CSF IL-6 elevation is indeed attributed to NPSLE.

6.4.5 Granulocyte-Colony Stimulating Factors

Recently the results in comparison of granulocyte-colony stimulating factors (G-CSF) levels in the CSF and serum samples of 30 patients with NPSLE, in whom the CSF and serum samples were obtained at the same time, suggested that in the patients with NPSLE the intrathecal concentrations of G-CSF were not influenced by the serum concentrations, indicating that production of G-CSF might take place in the CNS [10]. Furthermore, the mean level of CSF G-CSF was significantly higher in 30 patients with NPSLE than that in 22 patients with non-NPSLE [10].

Recently CSF G-CSF levels have been reported to be significantly higher in patients with neuromyelitis optica than in patients with other non-inflammatory neurological diseases [36]. G-CSF has been shown to be released from ECs [29] and to pass across the intact BBB [37]. Besides its role in hematopoiesis, G-CSF could also act as a neurotrophic factor, inducing neuorogenesis, as well as a protein to counteract apoptosis. These properties play a major role in the development of treatments for neurological diseases such as cerebral ischemia [37, 38]. Taken together, it is suggested that G-CSF might act to treat the damaged CNS intrathecally in patients with NPSLE.

6.5 Chemokines

Chemokines are chemoattractant cytokines which play key roles in the accumulation of inflammatory cells at the site of inflammation. Chemokines in humans comprise more than 50 small (8-to-10-kDa) heparin-binding proteins with 20–70 percent homology in amino acid sequences. Chemokines were originally identified by their chemotactic activity on bone marrow–derived cells [39, 40]. They are classified into at least four families according to the location of their cysteine residues. The four chemokine groups are CC, C, CXC, and CX3C, where C is a cysteine and X is any amino-acid residue, and their receptors are consequently classified as CCR, CR, CXCR, and CX3CR. The chemokine receptors are bound to the cell membrane through seven transmembrane helical segments coupled with a G-protein which transduces the intracellular signal. The two major subclasses include the CC chemokines where the cysteines are neighboring and the CXC chemokines where the cysteines are separated by one amino acid. The CXC chemokines mainly act on neutrophils and lymphocytes, whereas the CC chemokines mainly act on monocytes and lymphocytes without affecting neutrophils [41]. Fractalkine, in the CX3C family, is a cell-surface-bound protein, in which the first two cysteine residues are

Table 6.2 Chemokines in CSF of NPSLE

Chemokines	NPSLE	Control group	Authors	Year
MCP-1/CCL2	Increased	Non-NPSLE	Iikuni et al. [7]	2006
	Increased	Non-NPSLE	Fragoso-Loyo et al. [19]	2007
	Increased	Non-NPSLE	Yoshio et al. [10]	2016
RANTES/CCL5	Increased	Non-NPSLE	Fragoso-Loyo et al. [19]	2007
	Increased	HI, neurocysticerosis	Trysberg et al. [5]	2009
IL-8/CXCL8	Increased	Non-NPSLE, SM, non-AID	Fragoso-Loyo et al. [19]	2007
	Increased	MS, other AID	Santer et al. [24]	2009
	Increased	Non-NPSLE	Yoshio et al. [10]	2016
IP-10/CXCL10	Increased	Non-NPSLE	Okamoto et al. [9]	2004
	Increased	Non-NPSLE	Yoshio et al. [10]	2016
Fractalkine	Increased	Non-NPSLE	Yajima et al. [43]	2005
	Same	Non-NPSLE	Sato et al. [44]	2006
IP-10/MCP-1 ratio	Increased	Non-NPSLE	Okamoto et al. [8]	2006

AID autoimmune diseases, *HI* healthy individuals, *MS* multiple sclerosis, *SM* septic meningitis

separated by three amino acids. Fractalkine has potent chemoattractant activity for T cells and monocytes [42]. One characteristic feature of chemokines is the redundancy of the system. Several chemokines bind to more than one receptor and the majority of chemokine receptors have multiple ligands leading to the generation of multiple pathways directing similar cellular responses.

Several chemokines such as IL-8/CXCL8, the IP-10/CXCL10, fractalkine/CXCL1, regulated upon activation, normal T-cell expressed and secreted (RANTES) /CCL5 and monocyte chemoattractant protein (MCP) -1/CCL2 as well as IP-10/MCP-1 ratios have been reported to be elevated in the CSF from patients with NPSLE (Table 6.2) [5, 7–10, 19, 24, 43, 44].

6.6 Chemokines as Biomarkers

In this section, the role and diagnostic tools of each chemokine as a biomarker in NPSLE are described.

6.6.1 Monocyte Chemoattractant Protein-1/CCL2 (A Ligand of CCR2)

MCP-1/CCL2 (a ligand of CCR2) can attract monocytes, T cells, NK cells, and basophils [7, 29]. It is a high-affinity ligand for the CCR2 chemokine receptor that is constitutively expressed in monocytes but is expressed on lymphocytes only after stimulation by IL-2. Expression of CCR2 on monocytes can be down-regulated by lipopolysaccharides.

Our group and others have reported that CSF MCP-1/CCL2 levels are higher in NPSLE patients than in non-NPSLE patients [7, 19]. In addition, we reported that levels of MCP-1/CCL2 decreased after immunosuppressive treatment. Furthermore, we compared the levels of MCP-1/CCL2 among various neuropsychiatric symptoms. However, due to the paucity of sample size in some patient groups, we were unable to conclude which type of symptom was associated with the increase of CSF MCP-1/CCL2 levels in our study [7].

Recently, a comparison of MCP-1/CCL2 levels in the CSF and serum samples of 30 patients with NPSLE, in whom the CSF and serum samples were obtained at the same time, suggested that in the patients with NPSLE the intrathecal concentration of MCP-1/CCL2 were not influenced by the serum concentrations, indicating that production of MCP-1/CCL2 might take place in the CNS [10]. Furthermore, the mean level of CSF MCP-1/CCL2 was significantly higher in 30 patients with NPSLE than that in 22 patients with non-NPSLE [10].

6.6.2 Regulated Upon Activation, Normal T-Cell Expressed and Secreted (RANTES)/CCL5 (A Ligand of CCR1, CCR3, and CCR5)

RANTES/CCL5 is another CC chemokine which attracts monocytes, memory T cells and NK cells and is implicated in the pathophysiology of SLE, rheumatoid arthritis (RA) and multiple sclerosis [45]. Chemokine receptor CCR5 is preferentially expressed on T helper 1 (Th1) lymphocytes and has been reported to have an important role in the pathogenesis of RA. It has been reported that systemic administration of a small molecular weight antagonist of CCR5, SCH-X, suppressed the development of collagen-induced arthritis in a monkey model of RA [46]. Our group also provided evidence showing that systemic administration of TAK-779, a small molecular weight nonpeptide compound, inhibits the development of adjuvant-induced arthritis in rats [47].

Two reports showed that CSF level of RANTES/CCL5 in patients with NPSLE were higher than those in patients with non-NPSLE [5, 19]. Although the mean CSF level of RANTES/CCL5 was significantly higher in 30 patients with NPSLE than those in 22 patients with non-NPSLE, CSF RANTES/CCL5 levels were 1/100 of serum RANTES/CCL5 levels in 30 patients with NPSLE [10]. It is uncertain whether RANTES/CCL5 in the CSF contributes to the pathogenesis and appearance of NPSLE in the patients with SLE and whether the increased levels of RANTES/CCL5 in the CSF are caused by NPSLE.

6.6.3 Interleukin-8/CXCL8 (A Ligand of CXCR1 and CXCR2)

IL-8/CXCL8 was the first chemokine identified to be involved in leukocyte chemotaxis such as polymorphonuclear neutrophils and specific T cells [48, 49]. There are several reports showing that IL-8/CXCL8 levels in the CSF are elevated in NPSLE [19, 24].

Recently our group reported that in NPSLE patients intrathecal concentrations of IL-8/CXCL8 were not influenced by their serum concentrations, indicating that production of IL-8 might take place in the CNS [10]. Furthermore, the mean level of CSF IL-8/CXCL8 was significantly higher in 30 patients with NPSLE as compared to that in the 22 patients with non-NPSLE [10].

Our group also demonstrated that anti-NR2 induced the *in vivo* activation of human ECs, resulting in the enhanced production of cytokines such as IL-6 and IL-8/CXCL8 as well as the up-regulated expression of adhesion molecules such as ELAM-1, ICAM-1 and VCAM-1 through the activation of NF-kB pathway [28] as described elsewhere in the Sect. 6.4.4.

6.6.4 Interferon-Gamma Inducible Protein-10/CXCL10 (A Ligand of CXCR3)

IP-10/CXCL10 is expressed and secreted by monocytes and fibroblasts following stimulation with IFN-γ [50]. IP-10/CXCL10 is a high-affinity ligand for the CXCR3 chemokine receptor which is mainly expressed on natural killer cells and activated T cells, especially on Th1 cells. The predominance of Th1 versus Th2 cells in NPSLE patients remains unresolved. Okamoto et al. and other investigators have reported that IP-10/CXCL10 was up-regulated in the CSF of NPSLE [9, 19].

Recently a comparison of IP-10/CXCL10 levels in the CSF and serum samples of 30 patients with NPSLE, in whom the CSF and serum samples were obtained at the same time, disclosed that in the patients with NPSLE the intrathecal concentration of IP-10/CXCL10 were not influenced by their serum concentrations, indicating that production of IP-10/CXCL10 might take place in the CNS [10]. Furthermore, the mean level of CSF IP-10/CXCL10 was significantly higher in 30 patients with NPSLE as compared to that in the 22 patients with non-NPSLE [10].

Furthermore, as mentioned in the previous Sect. 6.6.3, the increased IP-10/CXCL10 in the CSF of NPSLE might be derived from the activation of ECs of the BBB, neuronal or glial cells.

6.6.5 Fractalkine/CX3CL1 (a Ligand of CX3CR1d)

The C chemokine family is represented by two chemokines, lymphotactin/XCL1 and SCM-1β/XCL2, whereas the CX3C chemokine family contains only one member, called fractalkine/CX3CL1 [51].

Fractalkine/CX3CL1 is synthesized by EC as a type 1 transmembrane protein which is then cleaved by proteolysis, possibly mediated by TNF-α-converting enzyme and ADAM 10, thereby yielding the soluble form of Fractalkine/CX3CL1 (sFKN). Fractalkine/CX3CL1 binds to a receptor known as CXCR1 and signals via the G protein pathway in NK cells, macrophages and a certain proportion of T cells.

Fractalkine/CX3CL1 plays important roles in the pathogenesis of RA by attracting pro-inflammatory cells, such as activated macrophages and T cells [52].

There is a report showing that levels of sFKN/sCX3CL1 were elevated in the CSF of NPSLE. In this report, both serum and CSF sFKN/sCX3CL1 levels declined along with successful treatment [43]. However, our group did not find a significant increase of sFKN/sCX3CL1 in CSF from NPSLE patients when compared with that of non-NPSLE patients [44].

6.6.6 Ratio of Two Different Chemokine Levels (The IP-10/MCP-1 Ratio)

The IP-10/MCP-1 ratio was reported to be a useful marker to detect NPSLE [8]. In this study, the IP-10/MCP-1 ratio in the NPSLE group was significantly higher than that in the non-NPSLE group (P = 0.0000014). The discriminative ability (area under the curve) of various ratios between NPSLE and non-NPSLE on Receiver Operating Characteristic (ROC) curve analysis was 0.63111 (IP-10/CXCL10), 0.67626 (MCP-1/CCL2) and 0.82672 (IP-10/MCP-1 ratio). These results supported the conclusion that CSF IP-10/MCP-1 ratios are higher in NPSLE patients than in non-NPSLE patients and that this index is a useful diagnostic marker of NPSLE [8].

6.7 Cytokines and Chemokines as Pathogenic Factors

6.7.1 Cytokines as Pathogenic Factors

Although some cytokines and chemokines are important biomarkers of NPSLE, the mechanism for the elevated levels of cytokines and chemokines are thus far unknown. As SLE is an autoimmune disorder characterized by numerous autoantibodies, a pathogenetic role for autoantibodies is theoretically suspected. Immune complexes in SLE can stimulate IFN-α and there is strong evidence in humans and in mice that IFN-α can cause neuropsychiatric manifestations as described in the section of **6.4.3 Interferon-α**. Santer DM et al. used a bioassay containing plasmacytoid dendritic cells to demonstrate that CSF from patients with NPSLE induced significantly higher IFN-α production compared with CSF from patients with multiple sclerosis or other autoimmune disease controls [24]. In addition to IFN-α, immune complexes formed by CSF autoantibodies significantly increased levels of IP-10/CXCL, IL-8/CXCL8 and MCP-1/CCL2. From these results they proposed a two-step model of NPSLE whereby CSF autoantibodies bind to antigens released by neurocytotoxic antibodies or other brain cell injury, and the resulting immune complexes stimulate IFN-α and proinflammatory cytokines and chemokines [24]. Recently, our group showed that IgG anti-NR2 from SLE patients directly activated ECs through the activation of NF-κB signaling, resulting in the up-regulation of adhesion molecules and cytokine production [28]. Thus, it is evident that autoantibodies alone can induce the production of cytokines without forming immune complexes. Further immunological studies are expected to show how autoantibodies in SLE patients work to promote the cytokine storm associated with the pathophysiology of NPSLE.

6.7.2 Chemokines as Pathogenic Factors

Our group reported that CSF MCP-1/CCL2 and IP-10/CXCL10 levels are higher in NPSLE patients than in non-NPSLE patients, indicating possible involvement of these chemokines in the pathogenesis of NPSLE [8, 9]. The receptor of IP-10/CXCL10, CXCR3, is predominantly expressed on natural killer cells and activated T cells, especially Th1 cells. On the other hand, the receptor of MCP-1/CCL2, CCR2, is expressed not only on activated T cells and natural killer cells but also on monocytes, basophils, and dendritic cells. CD4+ T cells populations that upregulate expression of the transcription factor RORγt can be differentiated into IL-17 producing CD4+ T cells (Th17 cells) that differ in phenotype and function from Th1 or Th2 cells. Th17 cells are thought to protect against bacteria and fungi and these cells are also involved in the pathogenesis of autoimmune diseases [50]. Interestingly, CCR2 is expressed on a subpopulation of Th17 cells which produce a large amount of IL-17 but little IFN-γ [53]. These results thus implicate the differential contribution of both CXCR3 and CCR2 signaling in the pathogenesis of NPSLE, especially on effector T cells such as Th1, Th2, and Th17 cells.

6.8 Summary

Although a large number of studies have been performed, the precise pathophysiology of NPSLE is not completely understood. As we described here, various cytokines and chemokines are highly expressed in the brain of NPSLE patients, and it is believed that these small molecules have important roles in the pathogenesis of NPSLE. However, the molecular mechanisms by which these molecules work in the course of the development of NPSLE have not yet been completely revealed. Cytokines and chemokines are expressed by the stimulation of NF-κB signaling as well as by signal transduction pathways involving other transcription factors [54]. As mentioned above, our group showed that IgG anti-NR2 from SLE patients direct NF-κB signaling in ECs, resulting in the up-regulation of adhesion molecules and cytokine production [28]. Therefore, autoantibodies which are characteristic feature of SLE bind to corresponding autoantigens on the EC surface and these interactions may stimulate signaling cascades, resulting in the activation of certain transcription factors (Fig. 6.1). Activation of signal transduction pathways involving these transcription factors might activate transcription and expression of cytokines and

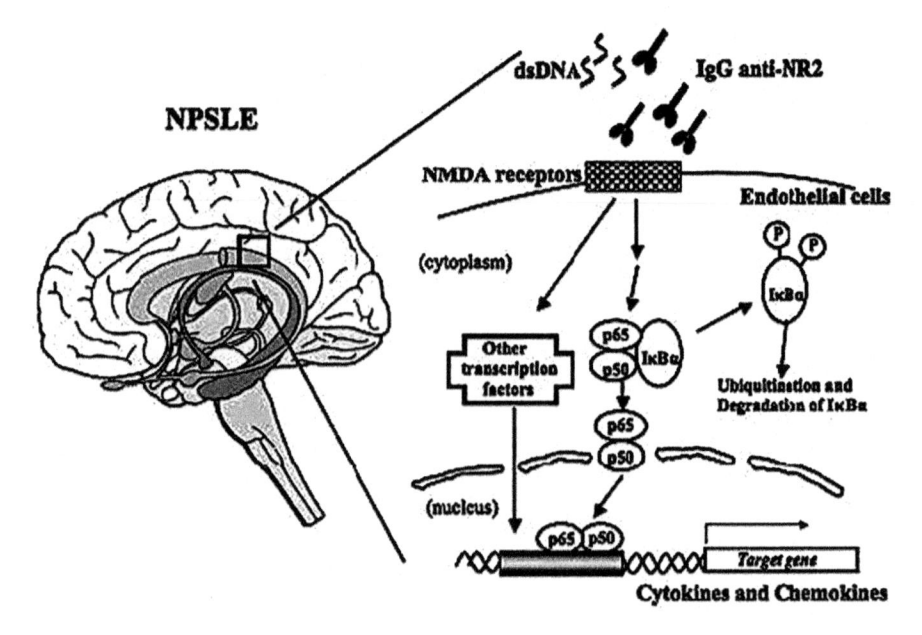

Fig. 6.1 Suggested mechanism of damages of central nervous system by anti-NMDAR NR2 antibodies in NPSLE

chemokines, resulting in a cytokine/chemokine storm and the development of NPSLE pathophysiology. Further molecular studies are required to prove this proposed mode of action for cytokines and chemokines. In addition, cytokines and chemokines are considered to be therapeutic targets of NPSLE. As most of the cytokines and chemokines involved in NPSLE have pleiotrophic roles in other biological processes, inhibition of these cytokines and chemokines might invite unexpected side effects *in vivo*. Therefore, cooperative contribution of both clinical studies and molecular biological studies is required for the development of ideal therapeutic strategies against NPSLE.

References

1. The American College of Rheumatology nomenclature and case definitions for neuropsychiatric lupus syndromes. ACR Ad Hoc Committee on Neuropsychiatric Lupus Nomenclature. Arthritis Rheum. 1999; 42:599–608.
2. Bresnihan B. CNS lupus. Clin Rheum Dis. 1982;8:183–95.
3. Brey RL, et al. Neuropsychiatric syndromes in lupus: prevalence using standardized definitions. Neurology. 2002;58:1214–20.
4. Ainiala H, et al. The prevalence of neuropsychiatric syndromes in systemic lupus erythematosus. Neurology. 2001;57:496–500.
5. Trysberg E, et al. Intrathecal cytokines in systemic lupus erythematosus with central nervous system involvement. Lupus. 2000;9:498–503.

6. Svenungsson E, et al. Increased levels of proinflammatory cytokines and nitric oxide metabolites in neuropsychiatric lupus erythematosus. Ann Rheum Dis. 2001;60:372–9.
7. Iikuni N, et al. Raised monocyte chemotactic protein-1 (MCP-1)/CCL2 in cerebrospinal fluid of patients with neuropsychiatric lupus. Ann Rheum Dis. 2006;65:253–6.
8. Okamoto H, et al. IP-10/MCP-1 ratio in CSF is a useful diagnostic marker of neuropsychiatric lupus patients. Rheumatology. 2006;45:232–4.
9. Okamoto H, et al. Interferon-inducible protein 10/CXCL10 is increased in the cerebrospinal fluid of patients with central nervous system lupus. Arthritis Rheum. 2004;50:3731–2.
10. Yoshio T, et al. IL-6, IL-8, IP-10, MCP-1 and C-CSF are significantly increased in cerebrospinal fluid but not in sera of patients with central neuropsychiatric lupus erythematosus. Lupus. 2016;25:997–1003.
11. Alcocer-Varela J, et al. Interleukin-1 and interleukin-6 activities are increased in the cerebrospinal fluid of patients with CNS lupus erythematosus and correlate with local late T-cell activation markers. Lupus. 1992;1:111–7.
12. Gilad R, et al. Cerebrospinal fluid soluble interleukin-2 receptor in cerebral lupus. Br J Rheumatol. 1997;36:190–3.
13. Jönsen A, et al. The heterogeneity of neuropsychiatric systemic lupus erythematosus is reflected in lack of association with cerebrospinal fluid cytokine profiles. Lupus. 2003;12:846–50.
14. Jara LJ, et al. Prolactin and interleukin-6 in neuropsychiatric lupus erythematosus. Clin Rheumatol. 1998;17:110–4.
15. Hirohata S, Miyamoto T. Elevated levels of interleukin-6 in cerebrospinal fluid from patients with systemic lupus erythematosus and central nervous system involvement. Arthritis Rheum. 1990;33:644–9.
16. Hirohata S, Hayakawa K. Enhanced interleukin-6 messenger RNA expression by neuronal cells in a patient with neuropsychiatric systemic lupus erythematosus. Arthritis Rheum. 1999;42:2729–30.
17. Tsai CY, et al. Cerebrospinal fluid interleukin-6, prostaglandin E2 and autoantibodies in patientswith neuropsychiatric systemic lupus erythematosus and central nervous system infections. Scand J Rheumatol. 1994;23:57–63.
18. Dellalibera-Joviliano R, et al. Kinins and cytokines in plasma and cerebrospinal fluid of patients with neuropsychiatric lupus. J Rheumatol. 2003;30:485–92.
19. Fragoso-Loyo H, et al. Interleukin-6 and chemokines in the neuropsychiatric manifestations of systemic lupus erythematosus. Arthritis Rheum. 2007;56:1242–50.
20. Winfield JB, et al. IntrathecalIgG synthesis and blood-brain barrier impairment in patients with systemic lupus erythematosus and central nervous system dysfunction. Am J Med. 1983;74:837–44.
21. Shiozawa S, et al. Interferon-alpha in lupus psychosis. Arthritis Rheum. 1992;35:417–22.
22. Aringer M, et al. Safety and efficacy of tumor necrosis factor alpha blockade in systemic lupus erythematosus: an open-label study. Arthritis Rheum. 2004;50:3161–9.
23. Rahman A, Isenberg DA. Systemic lupus erythematosus. N Engl J Med. 2008;358:929–39.
24. Santer DM, et al. Potent induction of IFN-alpha and chemokines by autoantibodies in the cerebrospinal fluid of patients with neuropsychiatric lupus. J Immunol. 2009;182:1192–201.
25. Wichers M, Maes M. The Psychoneuroimmuno-pathophysiology of cytokine-induced depression in humans. Int J Neuropsychopharmacol. 2002;5:375–88.
26. Prather AA, Rabinovitz M, Pollock BG, et al. Cytokine-induced depression during IFN- alpha treatment: the role of IL-6 and sleep quality. Brain Behav Immun. 2009;23:1109–16.
27. Hirohata S, et al. NPSLE Research Subcommittee. Accuracy of cerebrospinal fluid IL-6 testing for diagnosis of lupus psychosis. A multicenter retrospective study. Clin Rheumatol. 2009; 28:1319–23.
28. Yoshio T, et al. IgG anti-NR2 glutamate receptor autoantibodies from patients with systemic lupus erythematosus activate endothelial cells. Arthritis Rheum. 2013;65:457–63.
29. Lenhoff S, et al. Granulocyte interactions with GM-CSF and G-CSF secretion by endothelial cells and monocytes. Eur Cytokine Netw. 1999;10:525–32.

30. Cassatella MA, et al. Regulated production of the interferon-gamma-inducible protein-10 (IP-10) chemokine by human neutrophils. Eur J Immunol. 1997;27:111–5.
31. Deshmane SL, et al. Monocyte Chemoattractant Protein-1 (MCP-1): an overview. J Interf Cytokine Res. 2009;29:313–26.
32. Katsumata Y, et al. Diagnostic reliability of cerebral spinal fluid tests for acute confusional state (delirium) in patients with systemic lupus erythematosus: interleukin 6 (IL-6), IL-8, interferon-alpha, IgG index, and Q-albumin. J Rheumatol. 2007;34:2010–7.
33. George-Chandy A, et al. Raised intrathecal levels of APRIL and BAFF in patients with systemic lupus erythematosus: relationship to neuropsychiatric symptoms. Arthritis Res Ther. 2008;10:R97.
34. Hirohata S, et al. Clinical characteristics of neuro-Behcet's disease in Japan: a multicenter retrospective analysis. Mod Rheumatol. 2012;22:405–13.
35. Fujita Y, et al. Aseptic meningitis in mixed connective tissue disease: cytokine and anti-U1RNP antibodies in cerebrospinal fluids from two different cases. Mod Rheumatol. 2008;18:184–8.
36. Matsushita T, et al. Characteristic cerebrospinal fluid cytokine/chemokine profiles in neuromyelitis optica, relapsing remitting or primary progressive multiple sclerosis. PLoS One. 2013;8:e61835.
37. Schneider A, et al. The hematopoietic factor G-CSF is a neuronal ligand that counteracts programmed cell death and drives neurogenesis. J Clin Invest. 2005;115:2083–98.
38. Pitzer C, et al. Granulocyte-colony stimulating factor improves outcome in a mouse model of amyotrophic lateral sclerosis. Brain. 2008;131(Pt 12):3335–47.
39. Charo IF, Ransohoff RM. The many roles of chemokines and chemokine receptors in inflammation. N Engl J Med. 2006;354:610–21.
40. Iwamoto T, et al. Molecular aspects of rheumatoid arthritis: chemokines in the joints of patients. FEBS J. 2008;275:4448–55.
41. Luster GS. Mechanisms of disease: chemokines-chemotactic cytokines that mediate inflammation. N Engl J Med. 1998;338:436–45.
42. Bazan JF, et al. A new class of membrane-bound chemokine with a CX3C motif. Nature. 1997;385:640–4.
43. Yajima N, et al. Elevated levels of soluble fractalkine in active systemic lupus erythematosus: potential involvement in neuropsychiatric manifestations. Arthritis Rheum. 2005;52:1670–5.
44. Sato E, et al. Soluble fractalkine in the cerebrospinal fluid of patients with neuropsychiatric lupus. Ann Rheum Dis. 2006;65:1257–9.
45. Schall TJ, et al. Selective attraction of monocytes and T lymphocytes of the memory phenotype by cytokine RANTES. Nature. 1990;347:669–71.
46. Vierboom MP, et al. Inhibition of the development of collagen-induced arthritis in rhesus monkeys by a small molecular weight antagonist of CCR5. Arthritis Rheum. 2005;52:627–36.
47. Okamoto H, Kamatani N. A CCR-5 antagonist inhibits the development of adjuvant arthritis in rats. Rheumatology. 2006;45(2):230.
48. Moser B, Loetscher P. Lymphocyte traffic control by chemokines. Nat Immunol. 2001;2:123–8.
49. Baggiolini M, et al. Neutrophil-activating peptide-1/interleukin 8, a novel cytokine that activates neutrophils. J Clin Invest. 1989;84:1045–9.
50. Bromley SK, et al. Orchestrating the orchestrators: chemokines in control of T cell traffic. Nat Immunol. 2008;9:970–80.
51. Stievano L, et al. C and CX3C chemokines: cell sources and physiopathological implications. Crit Rev Immunol. 2004;24:205–28.
52. Murphy G, et al. Fractalkine in rheumatoid arthritis: a review to date. Rheumatology. 2008;47:1446–51.
53. Sato W, et al. Human Th17 cells are identified as bearing CCR2+CCR5- phenotype. J Immunol. 2007;178:7525–9.
54. Okamoto H, et al. Molecular aspects of rheumatoid arthritis: role of transcription factors. FEBS J. 2008;275:4463–70.

Chapter 7
Diagnosis and Differential Diagnosis

Taku Yoshio and Hiroshi Okamoto

Abstract Neuropsychiatric systemic lupus erythematosus (NPSLE) is a life-threatening disorder and early diagnosis and proper treatment are critical for the management of patients with this disease. NPSLE can manifest as a range of neurological and psychiatric features, which are classified using the ACR case definitions for 19 neuropsychiatric syndromes. Approximately one-third of all NPSLE events in patients with SLE are primary manifestations of SLE-related autoimmunity, with seizure disorders, cerebrovascular disease, acute confusional state and neuropathy being the most common. Such primary NPSLE events are a consequence either of autoantibodies and inflammatory mediators, or of microvasculopathy and thrombosis. Diagnosis of NPSLE requires the exclusion of other causes, and clinical assessment directs the selection of appropriate examinations. These examinations include measurement of autoantibodies, analysis of cerebrospinal fluid, electrophysiological studies, neuropsychological assessment and neuroimaging to evaluate brain structure and function. This chapter reviews the important key points for the correct diagnosis and the differential diagnosis.

Keywords NPSLE · Diagnosis · Differential diagnosis · Corticosteroid-induced psychiatric disorders (CIPD) · Autoantibodies · Cytokines · Chemokines

---

T. Yoshio (✉)
Division of Rheumatology and Clinical Immunology, School of Medicine, Jichi Medical University, Shimotsuke-shi, Tochigi, Japan
e-mail: takuyosh@jichi.ac.jp

H. Okamoto
Minami-otsuka institute of technology, Minami-otsuka Clinic, Tokyo, Japan

7.1 Introduction

Systemic lupus erythematosus (SLE) is a chronic multisystem inflammatory autoimmune disease with a waxing and waning course and a broad spectrum of clinical presentations [1]. The involvement of the nervous system in SLE patients leads to a nonspecific and heterogeneous group of neuropsychiatric manifestations [2]. A major issue in clinical evaluation is the attribution of neuropsychiatric symptoms to SLE. No laboratory or radiological biomarker nor other formal system exists for establishing a diagnosis in neuropsychiatric SLE (NPSLE). In clinical practice, an individual multidisciplinary diagnostic approach based on the suspected cause and severity of symptoms is recommended [3].

In this chapter, we describe the standard classification of NPSLE which was produced by the American College of Rheumatology (ACR) for the diagnosis of NPSLE, risk factors for NPSLE, SLE disease activity, clinical and laboratory examinations for diagnosis of NPSLE, diagnostic approach of NPSLE, guidelines for diagnosis of NPSLE and the important diseases that should be differentiated from NPSLE.

7.2 Classification of NPSLE

Many previous classifications of NPSLE lacked definitions of individual manifestations and standardization for investigation and diagnosis. In 1999, the ACR produced a standard nomenclature and set of case definitions for 19 neuropsychiatric syndromes known to occur in SLE (Table 7.1) [4]. These syndromes can be segregated into central and peripheral nervous system involvement [4], and diffuse and focal neuropsychiatric events [5]. The ACR classification is comprehensive in the scope of neuropsychiatric manifestations it describes, and provides guidance on investigations and diagnostic criteria for each. However, the classification has never intended to be specific for neuropsychiatric events caused exclusively by SLE. Thus, using the ACR classification in clinical practice it is important to attribute events to SLE and nonSLE causes to optimize the care of individual patients presenting with neuropsychiatric events.

7.3 Risk Factors for NPSLE

It is helpful for the diagnosis of NPSLE to bear in mind risk factors for various manifestations of NPSLE. Risk factors consistently associated with NPSLE events are shown as follows:

1. General SLE activity or damage, especially for seizure disorders and severe cognitive dysfunction [6–8].
2. Previous events or other concurrent NPSLE manifestations [9–11].

7 Diagnosis and Differential Diagnosis

Table 7.1 Neuropsychiatric syndromes observed in systemic lupus erythematosus

Central nervous system
Focal manifestations
Aseptic meningitis
Cerebrovascular disease
Demyelinating syndrome
Headache (including migraine and benign intracranial hypertension)
Movement disorder (chorea)
Myelopathy
Seizure disorders
Diffuse manifestations
Acute confusional state
Anxiety disorder
Cognitive dysfunction
Mood disorder
Psychosis
Peripheral nervous system
Acute inflammatory demyelinating polyradiculoneuropathy (Guillain-Barre syndrome)
Autonomic disorder
Mononeuropathy, single/multiplex
Myasthenia gravis
Neuropathy, cranial
Plexopathy
Polyneuropathy

Arthritis Rheum 42:599–608, 1999

3. Antiphospholipid antibodies (persistently positive moderate-to-high anticardiolipin or anti β_2-glycoprotein IgG/IgM titers or the lupus anticoagulant), especially for cerebrovascular disease (CVD) [7, 10], seizure disorder [6, 9], moderate-to-severe cognitive dysfunction [8, 12], myelopathy [13] and movement disorder [12].

7.4 SLE Disease Activity

Some studies [14–16], but not all [17, 18], have found an association between increased global SLE disease activity and neuropsychiatric events attributed to SLE. When the association between SLE disease activity index (SLEDAI) and the appearance of NPSLE were previously investigated, NPSLE occurred in the high scores of SLEDAI [19].

The evaluation of SLE disease activity, such as SLEDAI in organ systems other than neuropsychiatric events is important to diagnose NPSLE correctly and ensure appropriate management of neuropsychiatric events in patients with SLE. Such assessment might also help attribute these events to SLE and non-SLE causes. This association is probably more robust for diffuse rather than focal neuropsychiatric events.

7.5 Diagnostic Approach of NPSLE

Neuropsychiatric events may occur in patients when the presence of SLE or connective tissue disorders other than SLE is not confirmed. It is thus important to assess the presence or absence of SLE and other connective tissue disorders such as Sjogren's syndrome and mixed connective tissue disease. Of equally importance, the clinicians must realize that the presence of antinuclear antibody (ANA) in a patient with neurologic symptoms does not imply that the patient has NPSLE or, for that matter, SLE at all.

The evaluation of SLE patients with (new) signs or symptoms suggestive of NPSLE is comparable to that in non-SLE patients who present with the same neuropsychiatric manifestations [20]. Clinicians need to initially aim to exclude secondary causes such as infections, metabolic or endocrine disturbances and adverse drug reactions (Table 7.2).

The important diseases which should be differentiated from NPSLE are described in the Sect. 7.8.

Table 7.2 Secondary (non-lupus) causes of neuropsychiatric manifestations in systemic lupus erythematosus

Infection	
Medications	
Thrombotic thrombocytopenia purpura	
Hypertension	
Posterior reversible leukoencephalopathy syndrome	
Metabolic disturbance	Hyperglycemia or hypoglycemia
	Electrolyte imbabnces (Na^{+2}, Ca^{+2})
	Uremia
Hypoxemia	
Fever	
Thyroid disease	
Vitamin B12 deficiency	
Atherosclerosis strokes	
Subdural hematoma	
Berry aneurysm or cerebral hemorrhage	
Cerebral lymphoma	
Fibromyalgia	
Reactive depression	
Sleep apnea	
Other primary neurologic or psychiatric diseases	

7.6 Clinical and Laboratory Examination for the Diagnosis of NPSLE

No single test can diagnose NPSLE. After excluding secondary causes, the diagnosis of NPSLE can only be confirmed if a patient's neuropsychiatric symptoms can be corroborated with objective abnormalities in the neuropsychological examination, cerebrospinal fluid (CSF) examination, neuroimaging studies, electroencephalography, and/or biopsy. Therefore, a methodologic work-up is essential for the patient with SLE who complains of neuropsychiatric symptoms [21, 22].

A careful and thorough history taking and physical examination, including a complete neurologic and mental status (psychiatric) evaluation, must be performed on each patient.

7.6.1 Clinical and Laboratory Tests

In SLE patients who have neuropsychiatric symptoms, the clinical and laboratory tests which are necessary to confirm the diagnosis of NPSLE and to exclude other causes are shown in Table 7.3. A complete blood count and urinalysis should be obtained for disease activity and to rule out infection. If thrombocytopenia is present, the blood smear should be examined for schistocytes to exclude thrombotic thrombocytopenia purpura.

Blood chemistry tests, including electrolytes, creatinine, glucose, and liver-associated enzymes, are obtained to exclude metabolic abnormalities that can cause neurologic dysfunction. Complement (C3/C4, or CH50) determinations, anti-dsDNA antibodies and anti-Sm antibodies should be obtained to assess disease activity. The presence of antiphospholipid antibodies (lupus anticoagulant, anticardiolipin antibodies, anti-β_2 glycoprotein I antibodies) should be also determined.

Other tests for hypercoagulable states, including factor V Leiden, protein C and S levels, serum antithrombin III levels, and prothrombin 20210A mutation, may be indicated in selected patients. Most patients with SLE will have an elevated erythrocyte sedimentation rate and a normal or mildly elevated C-reactive protein. A significantly elevated C-reactive protein (>6 mg/dL) usually indicates systemic vasculitis or infection. A fasting lipid profile and homocysteine levels are obtained to establish vascular risk factors.

Table 7.3 Laboratory evaluation and diagnostic imaging of patients with systemic lupus erythematosus and neuropsychiatric manifestations

Complete blood count and peripheral blood smear	
Blood chemistry and serology	Electrolytes, creatinine, glucose, liver-associated enzymes
	C3/C4 and/or CH50
	Anti-dsDNA antibodies
	Anti-Sm antibodies
	Antiphospholipid antibodies (lupus anticoagulant, anticardiolipin, anti-β2 glycoprotein I)
Urinalysis	
Cerebrospinal fluid [CSF]	Cell count, protein, glucose, gram stain, other special stains including India ink (*Cryptococcus*), venereal disease research laboratory test and cultures (including polymerase chain reaction for herpes simplex virus, varicella zoster virus, and JC viruses).
	IgG levels, Q-albumin ratio, oligoclonal bands, IgG index
Brain and/or spinal cord magnetic resonance imaging [MRI]	T1/T2-weighted imaging
	A fluid-attenuated inversion recovery sequence [FLAIR]
	Diffusion weighted imaging [DWI]
	A gadolinium-enhanced T1-weighted sequence
Electroencephalography	
Other tests	C-reactive protein
	Serum and CSF antineuronal antibodies
	Serum and CSF neuromyelitis optica [NMO] IgG/ anti-aquaporin 4 antibodies
	Serum and CSF anti-ribosomal P protein antibodies
	Serum and CSF anti-NR2 glutamate receptor antibodies
	Computed tomography [CT] of brain
	CT or magnetic resonance angiogram [MRA]
	Cerebral angiography
	Single photon emission computed tomography [SPECT]
	Positron emission tomography
	18F–fluro-d-glucose positron emission tomography [FDG-PET]

7.6.2 Autoantibodies

More than 20 autoantibodies in the serum and CSF have been reported to be associated with NPSLE [23–25]. They have been detected by a variety of methods using multiple different substrates. Over one half of them are autoantibodies that react to brain antigens, whereas the remaining are systemic autoantibodies. However, many of these autoantibodies are not routinely available and remain investigational.

Notably, the five autoantibodies that are clinically available (antiphospholipid, anti-ribosomal P protein, antineuronal, anti-NR2 glutamate receptor, and neuromyelitis optica (NMO) IgG/anti-aquaporin 4 antibodies) deserve further attention.

7.6.3 CSF Tests

CSF analysis is useful in all patients with SLE who have had a change in neurologic status, particularly to exclude infection or other secondary causes of CNS dysfunction. In patients with NPSLE, CSF results may be unremarkable (50%). However, patients with NPSLE may have such abnormalities that are helpful in confirming the diagnosis and guiding management. Consensus panels recommend that routine CSF tests, IgG index, and oligoclonal bands be determined on all patients suspected of having NPSLE [4, 21].

7.6.3.1 Routine CSF Tests

Routine CSF tests include cell count with differentiation, protein, glucose, Gram stain, other special stains including India ink (*Cryptococcus*), venereal disease research laboratory test and cultures (including polymerase chain reaction for herpes simplex virus, varicella zoster virus, and JC viruses, if indicated).

Pleocytosis (more than 100 cells per high-power field) and elevated protein (70–110 mg/dL) are found in some patients with active NPSLE. Protein abnormalities are common (22% to 50%) than pleocytosis (6% to 34%) [26]. Neutrophilic pleocytosis with elevated protein suggests cerebral vasculitis with ischemia if infection is ruled out. Patients with antiphospholipid antibodies and neurologic thromboembolic events frequently have elevated protein levels with mild or no pleocytosis.

The CSF glucose level is rarely (3% to 8%) decreased (30 to 40 mg/dL) in NPSLE. CSF pleocytosis, elevated protein levels, and low glucose should always raise suspicion of an acute or chronic infection before attributing these abnormalities to NPSLE.

7.6.3.2 CSF Immunologic Tests

CSF IgG levels are elevated in 69% to 96% of patients with NPSLE, and a level greater than 6 mg/dL almost always indicates NPSLE, although it is present in only 40% of patients with NPSLE. An elevated CSF Q-albumin ratio, indicating a break in the blood brain barrier, has been noted in up to one third of patients, especially those with progressive encephalopathy, transverse myelitis, and strokes [22, 26]. Several groups have now confirmed that an elevated IgG index, the presence of oligoclonal bands or both are observed in up to 80% of patients, particularly in those with diffuse manifestations, such as encephalopathy and psychosis [22, 26, 27]. Patients with focal manifestations, such as stroke due to antiphospholipid antibodies, typically do not have an elevated IgG index or oligoclonal bands, unless they also have a coexistent encephalopathy (complex presentation) [22]. These abnormalities have been shown to normalize in some patients after successful therapy [22, 27].

7.6.3.3 CSF Autoantibodies

Using neuroblastoma cells as the antigen source, CSF levels of antineuronal antibodies were found to be significantly elevated in patients with lupus psychosis compared with those with nonpsychotic NPSLE or non-SLE controls [28].

Furthermore, 90% of the patients with diffuse manifestations of psychosis, encephalopathy or generalized seizures had elevated IgG antineuronal antibodies, compared with only 25% of patients with focal manifestations of hemiparesis or chorea. Notably the antineuronal antibodies were concentrated eightfold in the CSF, relative to its concentration in paired serum samples [29].

7.6.3.4 CSF Cytokine and Chemokine

Several cytokines (interleukin [IL]-6, interferon-α and granulocyte-colony stimulating factor) and chemokines (IL-8, interferon-γ-inducible-10, monocyte chemotactic protein-1) and matrix metalloproteinase-9 have been reported to be elevated in the CSF of patients with active NPSLE and may be important in the pathogenesis [30, 31]. Measurements of these mediators, especially IL-6, may be useful in the diagnosis and to monitor immunologic activity and neuronal damage. The intrathecal ratio of IP-10 to MCP-1 is significantly higher in patients with NPSLE than in patients with SLE without CNS symptoms. This IP-10/MCP-1 could be a useful marker of NPSLE [32, 33].

7.6.4 Neuroimaging Studies

Neuroimaging may detect NPSLE involvement and exclude other (neurosurgical, infectious) causes. The imaging technique of choice is magnetic resonance imaging (MRI) with T1/T2-weighted imaging, a fluid attenuating inversion recovery sequence, diffusion-weighted imaging (DWI) and a gadolinium-enhanced T1-weighted sequence. The average sensitivity of MRI in active NPSLE is 57% (64% in major vs 30% in minor NPSLE, 76% in focal vs 51% in diffuse NPSLE). The most frequent pathological pattern is small punctate hyperintensity focal lesions on T2-weighted images in subcortical and periventricular white matter, usually in the frontal-parietal regions. Unfortunately, these MRI lesions are also present in many patients without neuropsychiatric manifestations (specificity 60–82%) [34–36].

When conventional MRI is normal or does not provide an explanation for the signs and symptoms, advanced neuroimaging may be performed. Modalities to be considered (based on availability and local expertise) include quantitative MRI (magnetic resonance spectroscopy [37, 38], magnetisation transfer imaging [39, 40], diffusion tensor MRI [41], perfusion-weighted imaging) or radionuclide brain scanning (single photon emission computed tomography (SPECT) [42, 43], or positron emission tomography [44]. These imaging studies may reveal additional white matter and grey matter abnormalities, which, however, have modest specificity for NPSLE.

7.6.5 Electroencephalography

Conventional electroencephalography (EEG) is abnormal in 60% to 91% of adult and pediatric patients with NPSLE [26]. The most common finding is diffuse slowing with increased beta and delta background activity. Focal abnormalities and seizure activity can also be seen. Unfortunately, the EEG findings are not specific for NPSLE, and other disorders, including metabolic encephalopathies and drug effects, can give similar findings. Furthermore, up to 50% of patients with SLE without active NPSLE can have abnormal EEG. Consequently, a single abnormal EEG has limited diagnostic value for NPSLE. On occasion, however, an EEG may be very helpful, revealing unsuspected seizure activity, which was not clinically apparent.

7.7 Guidelines for Diagnosis of NPSLE

The EULAR standing committee for clinical affairs developed the recommendations for the management of SLE with neuropsychiatric manifestations [21]. The guidelines for the diagnosis that this committee recommended are shown in Table 7.4 (A part of Table 7.3 EULAR recommendations for the management of NPSLE is cited and revised and a part of supplementary Table S2, available online only, is also added [21]). When the clinicians diagnose NPSLE in the patients with SLE who have neuropsychological symptoms, these guidelines may be useful for the diagnostic tools.

Furthermore, we show the important key points for the diagnosis of various neuropsychological syndromes, such as headaches, cerebrovascular disease, cognitive dysfunction, seizure disorders, movement disorders, acute confusional states, psychosis, myelopathy, cranial neuropathy and peripheral nervous system disorders.

Table 7.4 EULAR recommendations for the diagnosis of neuropsychiatric systemic lupus erythematosus

Statement
General NPSLE
NPSLE
Neuropsychiatric events may precede, coincide, or follow the diagnosis of SLE but commonly (50–60%) occur within the first year after SLE diagnosis, in the presence of generalized disease activity (40–50%).
Cumulative incidence
Common (5–15% cumulative incidence) manifestations include CVD and seizures; relatively uncommon (1–5%): Severe cognitive dysfunction, major depression, ACS and peripheral nervous disorders; rare (<1%) are psychosis, myelitis, chorea, cranial neuropathies and aseptic meningitis.
Risk factors
Strong (fivefold increase) risk factors consistently associated with primary NPSLE are generalized SLE activity, previous severe NPSLE manifestations (especially for cognitive dysfunction and seizures), and antiphospholipid antibodies (especially for CVD, seizures, chorea).
Diagnostic work-up
In SLE patients with new or unexplained symptoms or signs suggestive of neuropsychiatric disease, initial diagnostic work-up should be similar to that in non-SLE patients presenting with the same manifestations.
Depending upon the type of neuropsychiatric manifestation, this may include lumbar puncture and CSF analysis (primarily to exclude CNS infection), EEG, neuropsychological assessment of cognitive function, NCS, and neuroimaging (MRI) to assess brain structure and function.
The recommended MRI protocol (brain and spinal cord) includes conventional MRI sequences (T1/T2, FLAIR), DWI, and gadolinium-enhanced T1 sequences.
Specific NPSLE disorders
CVD
Atherosclerotic/thrombotic/embolic CVD is common, hemorrhagic stroke is rare, and stroke caused by vasculitis is very rare in SLE patients; accordingly, immunosuppressive therapy is rarely indicated
Cognitive dysfunction
More common in Caucasians (10–20%) than in Asian (1–2%).
Mild or moderate cognitive dysfunction is common in SLE but severe cognitive impairment resulting in functional compromise is relatively uncommon (3–5%) and should be confirmed by neuropsychological tests in collaboration with a clinical neuropsychologist when available.
Seizure disorder
Single seizures are common in SLE patients and have been related to disease activity. Chance of recurrence is comparable to that in the general population.
The diagnostic work-up aims to exclude structural brain disease and inflammatory or metabolic conditions and includes MRI and EEG.
Acute confusional state (ACS)
Rates ranging 1.8–4.7% (including 'organic brain syndrome' cases).
Often in presence of generalized disease activity.
Type: hypo- or hyper-aroused states, ranging from delirium to coma.

(continued)

7 Diagnosis and Differential Diagnosis   103

Table 7.4 (continued)

Statement
In Japan the frequency of ACS is highest among diffuse psychiatric symptoms.
Lumbar puncture for CSF analysis and MRI should be considered to exclude non-SLE causes, especially infection.
The measurement of CSF IL-6 may be useful for the diagnosis of ACS, because the elevated levels in CSF IL-6 have been reported.
Major depression and psychosis
Major depression attributed to SLE alone is relatively uncommon while psychosis is rare; although steroid-induced psychosis may occur this is very rare.
There is no strong evidence to support the diagnostic utility of serological markers or brain imaging in major depression.
Myelopathy
Type: acute transverse myelopathy (most common), longitudinal myelopathy (>4 spinal cord segments affected, continuous or separate).
The diagnostic work-up includes gadolinium-enhanced MRI and CSF analysis.
Optic neuritis is commonly bilateral in SLE
The diagnostic work-up should include a complete ophthalmological evaluation (including funduscopy and fluoroangiography), MRI and visual evoked potentials.
Optic neuritis needs to be distinguished from ischemic optic neuropathy, which is usually unilateral, especially in patients with antiphospholipid antibodies.
Peripheral neuropathy
Peripheral neuropathy often co-exists with other neuropsychiatric manifestations and is diagnosed with electromyography and NCS.

ACS acute confusional state, *CNS* central nervous system, *CSF* cerebrospinal fluid, *CVD* cerebrovascular disease, *DWI* diffusion-weighted imaging, *EEG* electroencephagraphy, *FLAIR* fluid-attenuating inversion recovery sequence, *IL-6* interleukin-6, *MRI* magnetic resonance imaging, *NCS* nerve conduction studies, *NPSLE* neuropsychiatric systemic lupus erythematosus, *SLE* systemic lupus erythematosus, *T1/T2* T1/T2-weighted imaging

7.7.1 Headache

As the definition of lupus headaches five types of migraine, tension, cluster, headache from intracranial hypertension, and non-specific intractable headache are shown by the ACR [4]. Fragoso-Loyo H et al. have proposed that headache from intracranial hypertension and intractable non-specific headache are of an inflammatory nature and should remain as NPSLE syndromes, however, migraine is non-inflammatory and might be excluded from this nomenclature [45].

Although headache is frequently reported by SLE patients, several studies and a meta-analysis of epidemiological data found no evidence of an increased prevalence or a unique type of headache in SLE [46]. It is necessary to exclude aseptic or septic meningitis, sinus thrombosis (especially in patients with antiphospholipid antibodies), cerebral or subarachnoid hemorrhage. In the absence of high-risk features from the medical history and the physical examination (including fever or concomitant infection, immunosuppression, presence of antiphospholipid antibodies, use of anticoagulants, focal neurological signs, altered mental status, meningismus and generalized SLE activity), headache alone in an SLE patient requires no further investigation beyond the evaluation, if any, that would have been performed for non-SLE patients.

7.7.2 Cerebrovascular Disease

Ischemic stroke and/or TIA comprise over 80% of cerebrovascular disease (CVD) cases, whereas central nervous system (CNS) vasculitis is rare. CVD occurs commonly (50–60%) in the context of high disease activity and/or damage; other strong risk factors are persistently positive moderate-to-high titers of antiphospholipid antibodies, heart valve disease, systemic hypertension and old age.

In an acute stroke, MRI DWI excludes hemorrhage, assesses the degree of brain injury, and identifies the vascular lesion responsible for the ischemic deficit. Magnetic resonance angiography, angiography of computed tomography, or conventional angiography may help to characterize the vascular lesions and detect brain vasculature aneurysms in subarachnoid hemorrhage.

7.7.3 Cognitive Dysfunction

Most SLE patients have a mild-to-moderate degree of cognitive dysfunction with an overall benign course, and severe cognitive dysfunction develops only in 3–5% [47, 48]. Most commonly affected domains are attention, visual memory, verbal memory, executive function and psychomotor speed.

ACR has proposed a 1 h battery of neuropsychological tests for diagnosing cognitive dysfunction in SLE (sensitivity 80%, specificity 81%) [4]. The computer-based automated neuropsychological assessment metrics system has also been used. Indications for brain MRI include the followings: age less than 60 years, rapid unexplained or moderate-to-severe cognitive decline, recent and significant head trauma, new onset of other neurological symptoms or signs, and development of cognitive dysfunction in the setting of immunosuppressive or antiplatelet/anticoagulation therapy. Cerebral atrophy, the number and size of white matter lesions, and cerebral infarcts have been correlated with the severity of cognitive dysfunction [47, 49–51].

7.7.4 Seizure Disorders

Most seizures in SLE represent single isolated events, whereas recurrent seizures (epilepsy) are less common (12–22%) but have a significant impact on morbidity and mortality. Patients can experience generalized tonic–clonic seizures (67–88%) or partial (complex) seizures.

EEG abnormalities are common (60–70%) in SLE patients with seizure disorder, but typical epileptiform EEG patterns are only present in 24–50% and are predictive of seizure recurrence (positive predictive value 73%, negative predictive value 79%) [52, 53]. MRI can identify structural lesions causally related to seizure disorder and may reveal abnormalities such as cerebral atrophy (40%) and white matter lesions (50–55%). CSF examination is only useful to exclude infection.

7.7.5 Movement Disorders

Chorea (irregular, involuntary and jerky movements involving any part of the body in random sequence) is the best documented movement disorder in SLE, and has been associated with antiphospholipid antibodies and/or antiphospholipid syndrome. Brain imaging should be considered when other focal neurological signs are present or secondary causes of chorea need to be excluded. Most patients (55–65%) experience a single episode of chorea that subsides within days to a few months.

7.7.6 Acute Confusional State

Acute confusional state (ACS) is characterized by acute onset, fluctuating level of consciousness with decreased attention. Patients should be extensively evaluated for underlying precipitating conditions, especially infections and metabolic disturbances. CSF examination is recommended to exclude CNS infection and EEG may help diagnose underlying seizure disorder. Brain imaging is indicated if the patient has focal neurological signs, history of head trauma or malignancy, fever, or when the initial diagnostic work-up has failed to reveal any obvious cause of the ACS. Brain SPECT is sensitive (93%) and may help monitor response to treatment [54].

7.7.7 Psychosis

Lupus psychosis is characterized by delusions (false beliefs refuted by objective evidence) or hallucinations (perceptions in the absence of external stimuli). Although antiribosomal P protein antibodies have been associated with psychosis in prospective studies [55, 56], a meta-analysis has reported limited diagnostic accuracy (sensitivity 25–27%, specificity 75–80%) [57].

Brain MRI has modest sensitivity (50–70%) and specificity (40–67%) for lupus psychosis, and should be considered when additional neurological symptoms or signs are present. Brain SPECT identifies perfusion deficits in severe cases (80–100%) and residual hypoperfusion during clinical remission correlates with future relapse [58].

7.7.8 Myelopathy

SLE myelopathy presents as rapidly evolving transverse myelitis but ischemic/thrombotic myelopathy can also occur. Patients may present with signs of grey matter (lower motor neuron) dysfunction (flaccidity and hyporeflexia) or signs of white matter (upper motor neuron) dysfunction (spasticity and hyperreflexia); the latter can be associated more with neuromyelitis optica (NMO) and antiphospholipid [59]. Other major NPSLE manifestations are present in one third of cases, with optic neuritis being the most common (21–48%). Contrast-enhanced spinal cord MRI is useful to exclude cord compression and to detect T2-weighted hyperintensity lesions (70–93%). The involvement of more than four spinal cord segments indicates longitudinal myelopathy. This finding may be further investigated with determination of serum NMO IgG/anti-aquaporin 4 antibodies, which help diagnose co-existing NMO [60]. Brain MRI should be performed when other NPSLE symptoms or signs co-exist and in the differential diagnosis of demyelinating disorders. Mild-to-moderate CSF abnormalities are common (50–70%) but non-specific, while microbiological studies are important to exclude infectious myelitis.

7.7.9 Cranial Neuropathy

Most frequent cranial neuropathies involve the eighth, the oculomotor (third, fourth and sixth), and less commonly the fifth and seventh cranial nerves. Other neurological conditions, such as brainstem stroke and meningitis, should be excluded. Optic neuropathy includes inflammatory optic neuritis and ischemic/thrombotic optic neuropathy. Fundoscopy may reveal optic disc edema (30–40%) and visual field examination may show central or arcuate defects. Visual-evoked potentials may detect bilateral optic nerve damage before it is clinically apparent. Fluoroangiography should be performed when vaso-occlusive retinopathy is suspected. Co-existing transverse myelitis or seizure disorder may suggest an underlying inflammatory basis, while optic neuropathy with an altitudinal field defect, associated with antiphospholipid antibodies, renders an ischemic/thrombotic mechanism more likely. The diagnosis is supported by contrast-enhanced MRI showing optic nerve enhancement in 60–70%, while brain MRI abnormalities are also common (67%).

7.7.10 Peripheral Nervous System Disorders

Peripheral nervous system disorders include polyneuropathy (2–3%) and less commonly mononeuropathy (single, multiplex), acute inflammatory demyelinating polyradiculoneuropathy, myasthenia gravis, plexopathy, and present with altered sensation, pain, muscle weakness or atrophy. CNS involvement should be excluded by neuroimaging when focal neurological signs, gait disturbance, visual or urinary disorder, increased tendon reflexes and/or muscle tone are present. Nerve conduction studies and needle electromyography can identify mononeuropathies, differentiate multiple mononeuropathy versus polyneuropathy and distinguish axonal neuropathies from demyelinating neuropathies. CSF analysis is useful for diagnosis of inflammatory demyelinating polyradiculoneuropathy. Nerve biopsy is rarely needed to establish the diagnosis. If electrodiagnostic studies are normal, small-fiber neuropathy may be diagnosed by skin biopsy demonstrating loss of intraepidermal nerve fibers [61].

7.8 The Important Diseases for Differential Diagnosis

In this section, the diseases that should be differentiated from NPSLE are described.

7.8.1 Psychiatric Manifestations after Steroid Therapy

When the new-onset psychiatric manifestations appear in patients with SLE after the initiation of corticosteroid therapy (the dose of prednisone 1 mg/kg or more) or the increase of corticosteroid therapy, such as pulse intravenous methylprednisolone (1 g/day for 3 days), it is very difficult to determine whether these psychiatric manifestations are caused by SLE itself (NPSLE) or induced by steroids (corticosteroid-induced psychiatric disorders [CIPD]) [62]. CIPD occurs in 10% of patients treated with prednisone 1 mg/kg or more and it manifests primarily as mood disorder, such as manic or depressive state (93%), rather than psychosis [63].

It has been reported that CSF IL-6 levels are increased in patients with NPSLE, but not in SLE patients without NPSLE or with CIPD. Thus, the measurement of CSF IL-6 is useful for the differential diagnosis between NPSLE and CIPD [64]. The corticosteroid therapy may deteriorate psychiatric symptoms by reducing the brain blood flow, leading to the development of CIPD. The physicians should not reduce the steroid dose or cease the steroid therapy in case of CIPD in order to avoid the exacerbation of SLE disease activity.

7.8.2 Neuromyelitis Optica

Neuromyelitis optica (NMO), also known as Devic syndrome, is a severe demyelinating disorder of the CNS that causes longitudinal transverse myelitis of at least three vertebral segments and recurrent optic neuritis. NMO has been reported in patients with SLE [65], which is associated with NMO-specific autoantibodies whose antigenic target is aquaporin 4 [66], the most abundant water channel in the CNS [67]. Although NMO is a rare clinical presentation, suspicion of this syndrome in a patient with SLE warrants the measurement of IgG anti-aquaporin 4 antibodies.

7.8.3 Reversible Posterior Leukoencephalopathy Syndrome

Over the past decade, reversible posterior leukoencephalopathy syndrome (RPLS) has been recognized as an important secondary cause of neurologic dysfunction [68]. At onset, patients with SLE typically have seizures (75% to 100%), accelerated hypertension (90% to 95%), acute renal failure (85% to 90%), headache (70%), blurred vision (45% to 50%), and/or cortical blindness (30%). Notably, over 75% have had augmentation of their immunosuppressants (intravenous methylprednisolone, intravenous cyclophosphamide) within an average of 7 days before the development of RPLS. The majority (61%) have evidence of brain MRI abnormalities involving the posterior lobe circulation caused by vasogenic edema. Therapy includes prompt control of the blood pressure. Further increase in immunosuppressive therapy is contraindicated and potentially detrimental. Long-term anticonvulsant use is rarely needed once neuroimaging abnormalities resolve after an average of 25 days. With early recognition and prompt therapy, full neurologic recovery usually occurs.

7.8.4 Progressive Multifocal Leukoencephalopathy

Progressive multifocal leukoencephalopathy (PML) is a rare, deadly demyelinating disease of the CNS, which is caused by a reactivation of the DNA polyomavirus, the John Cunningham virus (JCV), and occurs in immunosuppressed hosts. Of note, most SLE patients who develop PML have been either subjected or are concomitantly under immunosuppressant therapy [69].

MRI is the most sensitive imaging method for the investigation of suspected PML, typical lesions appearing hyperintensity on FLAIR and T2-weighted sequences [70]. Isolation of the JCV in brain tissue confirms the diagnosis of PML. Polymerase chain reaction (PCR) analysis of CSF for the presence of JCV has also been proved useful in the diagnosis of PML [71, 72]. Typically the patients with PML present with cognitive impairment, altered mental status, aphasia, focal motor deficits, cortical blindness and behavioral changes [73, 74].

PML must be considered in the differential diagnosis of SLE patients presenting with unexplained neurologic symptoms or signs, and a low threshold for performing PCR analysis of CSF for JCV must be maintained. Furthermore, since negative PCR results do not exclude the diagnosis of PML, brain tissue biopsy should be considered in patients in whom clinical suspicion of PML remains high, despite negative results on PCR analysis of CSF for JCV.

7.9 Summary

Neuropsychiatric symptoms constitute an uncommon and poorly understood event in SLE patients, and pose a diagnostic challenge to the physician. Management of NPSLE patients has not evolved substantially in the last decades and is characterized by the lack of good evidence to date. It seems reasonable that increased understanding of the pathogenesis of NPSLE as well as the specific findings for NPSLE will promote the possibility of discovery of the diagnostic tools for the rapidly targeted therapy.

References

1. Tsokos GC. Systemic lupus erythematosus. N Engl J Med. 2011;365:2110–21.
2. Jeltsch-David H, Muller S. Neuropsychiatric systemic lupus erythematosus: pathogenesis and biomarkers. Nat Rev Neurol. 2014;10:579–96.
3. Zirkzee EJ, et al. Prospective study of clinical phenotypes in neuropsychiatric systemic lupus erythematosus; multidisciplinary approach to diagnosis and therapy. J Rheumatol. 2012;39:2118–26.
4. The American College of Rheumatology nomenclature and case definitions for neuropsychiatric lupus syndromes. Arthritis Rheum. 1999; 42:599–608.
5. Hanly JG, et al. Autoantibodies and neuropsychiatric events at the time of systemic lupus erythematosus diagnosis: results from an international inception cohort study. Arthritis Rheum. 2008;58:843–53.
6. Andrade RM, et al. Seizures in patients with systemic lupus erythematosus: data from LUMINA, a multiethnic cohort (LUMINA LIV). Ann Rheum Dis. 2008;67:829–34.
7. Mikdashi J, Handwerger B. Predictors of neuropsychiatric damage in systemic lupus erythematosus: data from the Maryland lupus cohort. Rheumatology. 2004;43:1555–60.
8. Tomietto P, et al. General and specific factors associated with severity of cognitive impairment in systemic lupus erythematosus. Arthritis Rheum. 2007;57:1461–72.
9. Appenzeller S, et al. Epileptic seizures in systemic lupus erythematosus. Neurology. 2004;63:1808–12.

10. Buján S, et al. Contribution of the initial features of systemic lupus erythematosus to the clinical evolution and survival of a cohort of Mediterranean patients. Ann Rheum Dis. 2003;62:859–65.
11. Mikdashi J, et al. Factors at diagnosis predict subsequent occurrence of seizures in systemic lupus erythematosus. Neurology. 2005;64:2102–7.
12. Sanna G, et al. Neuropsychiatric manifestations in systemic lupus erythematosus: prevalence and association with antiphospholipid antibodies. J Rheumatol. 2003;30:985–92.
13. Mok MY, et al. Antiphospholipid antibody profiles and their clinical associations in Chinese patients with systemic lupus erythematosus. J Rheumatol. 2005;32:622–8.
14. Govoni M, et al. Factors and comorbidities associated with first neuropsychiatric event in systemic lupus erythematosus: does a risk profile exist? A large multicentre retrospective cross-sectional study on 959 Italian patients. Rheumatology (Oxford). 2012;51:157–68.
15. Florica B, et al. Peripheral neuropathy in patients with systemic lupus erythematosus. Semin Arthritis Rheum. 2011;41:203–11.
16. Toledano P, et al. Neuropsychiatric systemic lupus erythematosus: magnetic resonance imaging findings and correlation with clinical and immunological features. Autoimmun Rev. 2013;12:1166–70.
17. Jarpa E, et al. Common mental disorders and psychological distress in systemic lupus erythematosus are not associated with disease activity. Lupus. 2011;20:58–66.
18. Shimojima Y, et al. Relationship between clinical factors and neuropsychiatric manifestations in systemic lupus erythematosus. Clin Rheumatol. 2005;24:469–75.
19. Morrison E, et al. Neuropsychiatric systemic lupus erythematosus: association with global disease activity. Lupus. 2014;23:370–7.
20. Bertsias G, et al. EULAR recommendations for the management of systemic lupus erythematosus. Report of a task force of the EULAR Standing Committee for International Clinical Studies including therapeutics. Ann Rheum Dis. 2008;67:195–205.
21. Bertsias GK, et al. EULAR recommendations for the management of systemic lupus erythematosus with neuropsychiatric manifestations: report of a task force of the EULAR standing committee for clinical affairs. Ann Rheum Dis. 2010;69:195–205.
22. West SG, et al. Neuropsychiatric lupus erythematosus: a 10-year prospective study on the value of diagnostic tests. Am J Med. 1995;99:153–63.
23. Zandman-Goddard G, et al. Autoantibodies involved in neuropsychiatric SLE and antiphospholipid syndrome. Semin Arthritis Rheum. 2007;36:297–315.
24. Kimura A, et al. Antibodies in patients with neuropsychiatric systemic lupus erythematosus. Neurology. 2010;74:1372–9.
25. Hanley JG, et al. Autoantibodies and neuropsychiatric events at the time of systemic lupus erythematosus diagnosis: results from an international inception cohort study. Arthritis Rheum. 2008;58:843–53.
26. West S. The nervous system. In: Wallace DJ, Hahn BH, editors. Dubois' lupus erythematosus. 7th ed. Phiadelphia: Lippincott-Williams & Wilkins; 2007. p. 707–46.
27. Hirohata S, Taketani TA. Serial study of changes in intrathecal immunoglobulin synthesis in a patient with central nervous system systemic lupus erythematosus. J Rheumatol. 1987;14:1055–7.
28. Isshi K, Hirohata S. Differential roles of the anti-ribosomal P antibody and antineuronal antibody in the pathogenesis of central nervous system involvement in systemic lupus erythematosus. Arthritis Rheum. 1998;41:1819–27.
29. Bluestein HG, et al. Cerebrospinal fluid antibodies to neuronal cells: association with neuropsychiatric manifestations of systemic lupus erythematosus. Am J Med. 1981;70:240–6.
30. Okamoto H, et al. Cytokines and chemokines in neuropsychiatric syndromes of systemic lupus erythematosus. J Biomed Biotechnol. 2010;2010:268436.
31. Yoshio T, et al. IL-6, IL-8, IP-10, MCP-1 and G-CSF are significantly increased in cerebrospinal fluid but not in sera of patients with central neuropsychiatric lupus erythematosus. Lupus. 2016;25:997–1003.

32. Aranow C, et al. Pathogenesis of the nervous system. In: Wallace DJ, Hahn BH, editors. Dubois' lupus erythematosus. 8th ed. Philadelphia: Lippincott-Williams & Wilkins; 2013. p. 363–7.
33. Okamoto H, et al. IP-10/MCP-1 ratio in CSF is an useful diagnostic marker of neuropsychiatric lupus patients. Rheumatology. 2006;45:232–4.
34. Kozora E, et al. Magnetic resonance imaging abnormalities and cognitive deficits in systemic lupus erythematosus patients without overt central nervous system disease. Arthritis Rheum. 1998;41:41–7.
35. Nomura K, et al. Asymptomatic cerebrovascular lesions detected by magnetic resonance imaging in patients with systemic lupus erythematosus lacking a history of neuropsychiatric events. Intern Med. 1999;38:785–95.
36. Sibbitt WL, et al. Magnetic resonance and computed tomographic imaging in the evaluation of acute neuropsychiatric disease in systemiclupus erythematosus. Ann Rheum Dis. 1989;48:1014–22.
37. Appenzeller S, et al. Evidence of reversible axonal dysfunction in systemic lupus erythematosus: a proton MRS study. Brain. 2005;128(Pt 12):2933–40.
38. Axford JS, et al. Sensitivity of quantitative (1) H magnetic resonance spectroscopy of the brain in detecting early neuronal damage in systemic lupus erythematosus. Ann Rheum Dis. 2001;60:106–11.
39. Bosma GP, et al. Multisequence magnetic resonance imaging study of neuropsychiatric systemic lupus erythematosus. Arthritis Rheum. 2004;50:3195–202.
40. Steens SC, et al. Selective gray matter damage in neuropsychiatric lupus. Arthritis Rheum. 2004;50:2877–81.
41. Hughes M, et al. Diffusion tensor imaging in patients with acute onset of neuropsychiatric systemic lupus erythematosus: a prospective study of apparent diffusion coefficient, fractional anisotropy values, and eigenvalues in different regions of the brain. Acta Radiol. 2007;48:213–22.
42. Castellino G, et al. Single photon emission computed tomography and magnetic resonance imaging evaluation in SLE patients with and without neuropsychiatric involvement. Rheumatology. 2008;47:319–23.
43. Zhang X, et al. Diagnostic value of single-photon-emission computed tomography in severe central nervous system involvement of systemic lupus erythematosus: a case-control study. Arthritis Rheum. 2005;53:845–9.
44. Weiner SM, et al. Diagnosis and monitoring of central nervous system involvement in systemic lupus erythematosus: value of F-18 fluorodeoxyglucose PET. Ann Rheum Dis. 2000;59:377–85.
45. Fragoso-Loyo H, et al. Inflammatory profile in cerebrospinal fluid of patients with headache as a manifestation of neuropsychiatric systemic lupus erythematosus. Rheumatology. 2013;52:2218–22.
46. Mitsikostas DD, et al. A meta-analysis for headache in systemic lupus erythematosus: the evidence and the myth. Brain. 2004;127(Pt 5):1200–9.
47. Tomietto P, et al. General and specific factors associated with severity of cognitive impairment in systemic lupus erythematosus. Arthritis Rheum. 2007;57:1461–72.
48. Panopalis P, et al. Impact of memory impairment on employment status in persons with systemic lupus erythematosus. Arthritis Rheum. 2007;57:1453–60.
49. Ainiala H, et al. Cerebral MRI abnormalities and their association with neuropsychiatric manifestations in SLE: a population-based study. Scand J Rheumatol. 2005;34:376–82.
50. Lapteva L, et al. Anti-N-methyl-d-aspartate receptor antibodies, cognitive dysfunction, and depression in systemic lupus erythematosus. Arthritis Rheum. 2006;54:2505–14.
51. Waterloo K, et al. Neuropsychological dysfunction in systemic lupus erythematosus is not associated with changes in cerebral blood flow. J Neurol. 2001;248:595–602.
52. Appenzeller S, et al. Epileptic seizures in systemic lupus erythema- tosus. Neurology. 2004;63:1808–12.

53. González-Duarte A, et al. Clinical description of seizures in patients with systemic lupus erythematosus. Eur Neurol. 2008;59:320–3.
54. Tokunaga M, et al. Efficacy of rituximab (anti-CD20) for refractory systemic lupus erythematosus involving the central nervous system. Ann Rheum Dis. 2007;66:470–5.
55. Briani C, et al. Neurolupus is associated with anti-ribosomal P protein antibodies: an inception cohort study. J Autoimmun. 2009;32:79–84.
56. Hanly JG, et al. Autoantibodies and neuropsychiatric events at the time of systemic lupus erythematosus diagnosis: results from an international inception cohort study. Arthritis Rheum. 2008;58:843–53.
57. Karassa FB, et al. Accuracy of anti-ribosomal P protein antibody testing for the diagnosis of neuropsychiatric systemic lupus erythematosus: an international meta-analysis. Arthritis Rheum. 2006;54:312–24.
58. Kodama K, et al. Single photon emission computed tomography in systemic lupus erythematosus with psychiatric symptoms. J Neurol Neurosurg Psychiatry. 1995;58:307–11.
59. Birnbaum J, et al. Distinct subtypes of myelitis in systemic lupus erythematosus. Arthritis Rheum. 2009;60:3378–87.
60. Pittock SJ, et al. Neuromyelitis optica and non organ-specific autoimmunity. Arch Neurol. 2008;65:78–83.
61. Tseng MT, et al. Skin denervation and cutaneous vasculitis in systemic lupus erythematosus. Brain. 2006;129(Pt 4):977–85.
62. Chau SY, Mok CC. Factors predictive of corticosteroid psychosis in patients with systemic lupus erythematosus. Neurology. 2003;61:104–7.
63. Nishimura K, et al. New-onset psychiatric disorders after corticosteroid therapy in systemic lupus erythematosus: an observational case-series study. J Neurol. 2014;261:2150–8.
64. Katsumata Y, et al. Diagnostic reliability of cerebral spinal fluid tests for acute confusional state (delirium) in patients with systemic lupus erythematosus: interleukin 6 (IL-6), IL-8, interferon-alpha, IgG index, and Q-albumin. J Rheumatol. 2007;34:2010–7.
65. Birnbaum J, Kerr D. Devic's syndrome in a woman with systemic lupus erythematosus: diagnostic and therapeutic implications of testing for the neuromyelitis optica IgG autoantibody. Arthritis Rheum. 2007;57:347–51.
66. Waters P, et al. Aquaporin-4 antibodies in neuromyelitis optica and longitudinally extensive transverse myelitis. Arch Neurol. 2008;65:913–9.
67. Lennon VA, et al. IgG marker of optic-spinal multiple sclerosis binds to the aquaporin-4 water channel. J Exp Med. 2005;202:473–7.
68. Budhoo A, Mody GM. The spectrum of posterior reversible encephalopathy in systemic lupus erythematosus. Clin Rheumatol. 2015;34:2127–34.
69. Calabrese LH, et al. M. Progressive multifocal leukoencephalopathy in rheumatic diseases: evolving clinical and pathologic patterns of disease. Arthritis Rheum. 2007;56:2116–28.
70. De Gascun CF, Carr MJ. Human polyomavirus r eactivation: disease pathogenesis and treatment approaches. Clin Dev Immunol. 2013;2013:373579.
71. Cinque P, et al. Progressive multifocal leukoencephalopathy in HIV-1 infection. Lancet Infect Dis. 2009;9:625–36.
72. d'Arminio Monforte A, et al. A comparison of brain biopsy and CSF-PCR in the diagnosis of CNS lesions in AIDS patients. J Neurol. 1997;244:35–9.
73. Garrels K, et al. Progressive multifocal leukoencephalopathy: clinical and MR response to treatment. AJNR Am J Neuroradiol. 1996;17:597–600.
74. Smith AB, et al. From the archives of the AFIP: central nervous system infections associated with human immunodeficiency virus infection: radiologic-pathologic correlation. Radiographics. 2008;28:2033–58.

Chapter 8
Imaging of Neuropsychiatric Systemic Lupus Erythematosus

Yoshiyuki Arinuma and Shunsei Hirohata

Abstract The diagnosis of neuropsychiatric manifestations in systemic lupus erythematosus (NPSLE) is challenging. Neuroimaging is very important technique for the evaluation of abnormalities occurred in the central nervous system (CNS). Computed tomography (CT) scan is one of the common techniques and is very useful to detect a large lesion such as ischemic stroke, hemorrhage and tumor, providing help to rule out CNS diseases other than SLE. Magnetic resonance imaging (MRI) is the best tool at present to detect parenchymal lesions in the CNS. Conventional MRI figures out exact abnormalities in the CNS causing neurologic or psychiatric symptoms. However, it should be remembered that MRI abnormalities in patients with NPSLE are not always specific for NPSLE. New MRI techniques can give us more detailed information in patients with NPSLE in addition to findings by conventional MRI. Functional analysis of the CNS by imaging system would be promising.

Keywords NPSLE · Neuroimaging · Central nervous system · MRI

Y. Arinuma (✉)
Department of Rheumatology and Infectious diseases, Kitasato University School of Medicine, Sagamihara, Kanagawa, Japan

S. Hirohata
Department of Rheumatology
Nobuhara Hospital, Tatsuno, Hyogo, Japan

Department of Rheumatology & Infectious Diseases
Kitasato University School of Medicine, Sagamihara, Kanagawa, Japan

© Springer International Publishing AG, part of Springer Nature 2018
S. Hirohata (ed.), *Neuropsychiatric Systemic Lupus Erythematosus*,
https://doi.org/10.1007/978-3-319-76496-2_8

8.1 Introduction

The diagnosis of neuropsychiatric syndromes in systemic lupus erythematosus (NPSLE) is still challenging [1], because NPSLE includes a variety of manifestations coming from heterogeneous etiologies, and the pathogenesis has not been sufficiently clear.

Neuroimaging examinations give us numerous information on the central nervous system (CNS) and its associated tissues surrounding CNS. Neuroimaging has actually 2 aspects regarding of their properties; the observation of the CNS structure and the evaluation of brain function. The structural abnormalities should be directly affected by the pathological changes of damaged sites. Certainly, the pattern of abnormal findings is reflecting a probable lesion such as infarction, hemorrhage, abscess, tumor etc., which can manifest focal sign. Occasionally, the structural abnormalities are associated with psychiatric manifestation. Some methodologies of neuroimaging are used to estimate a function of the brain, which can quantitatively measure blood flow, brain metabolism and biochemistry. For these reasons, use of neuroimaging has been found to be important for the diagnosis and management of NPSLE [2]. Although there are no specific abnormalities for NPSLE, evaluation with neuroimaging can support a diagnosis and management of NPSLE. In this chapter, characteristics of findings in each imaging modality are described. Thus, use of neuroimaging could be useful not only in neurologic syndromes (aseptic meningitis, cerebrovascular disease, demyelinating syndrome, headache, movement disorder, myelopathy, seizure disorder and cranial neuropathy), but in diffuse psychiatric/neuropsychological syndromes (acute confusional state, anxiety disorder, cognitive dysfunction, mood disorder and psychosis).

8.2 Computed Tomography (CT)

CT is one of the commonest devices and is especially powerful in emergency. CT is also so safe that there are few contraindications for most of the patients and it needs to spend only a moment to obtain images. CT is a very good tool for finding an acute CNS event such as hemorrhage as well as a wide range of infarction, and can help to quickly detect the presence of other diseases like brain abscess and tumor. Since CT is not useful to detect non-hemorrhagic lesions in brain tissue, it is usually performed in case of emergency to explore space occupying lesion or severe hemorrhage (Fig. 8.1). According to a recent nationwide population-based study, subarachnoid hemorrhage (SAH) has been indicated to be a complication of SLE with a high mortality rate [3]. Age, higher daily steroid use and a history of platelet or red blood cell transfusion need to be ruled out as risk factors for SAH in SLE. On the other hand, case series reported that patients with higher SLE disease activity had worse mortality due to SAH [4]. Brain vasculitis could be a pathogenetic factor of SAH, although it has not been clearly defined in NPSLE case definitions [5].

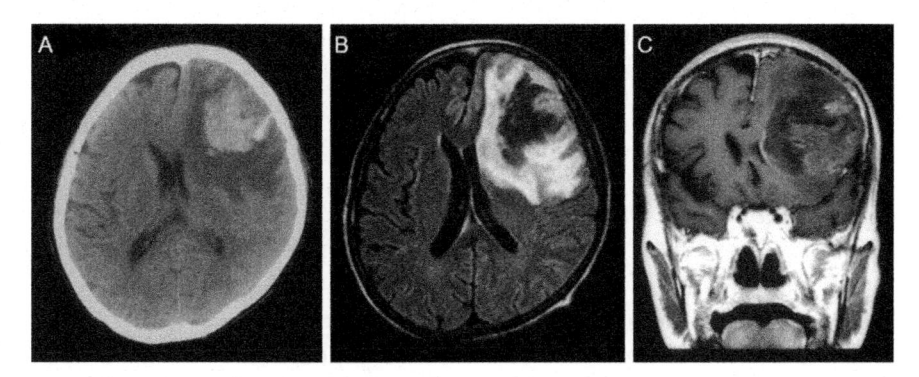

8 Imaging of Neuropsychiatric Systemic Lupus Erythematosus 115

Fig. 8.1 Hemorrhage and infarction in a patient with SLE with a large ischemic stroke due to antiphospholipid syndrome. Hemorrhagic lesion is shown at the center of the large infarcted area in left frontal lobe on CT scan (**a**). The wide and large fresh infarction with hematoma is observed in DWI (**b**) and T1-weighted image (**c**) on MRI

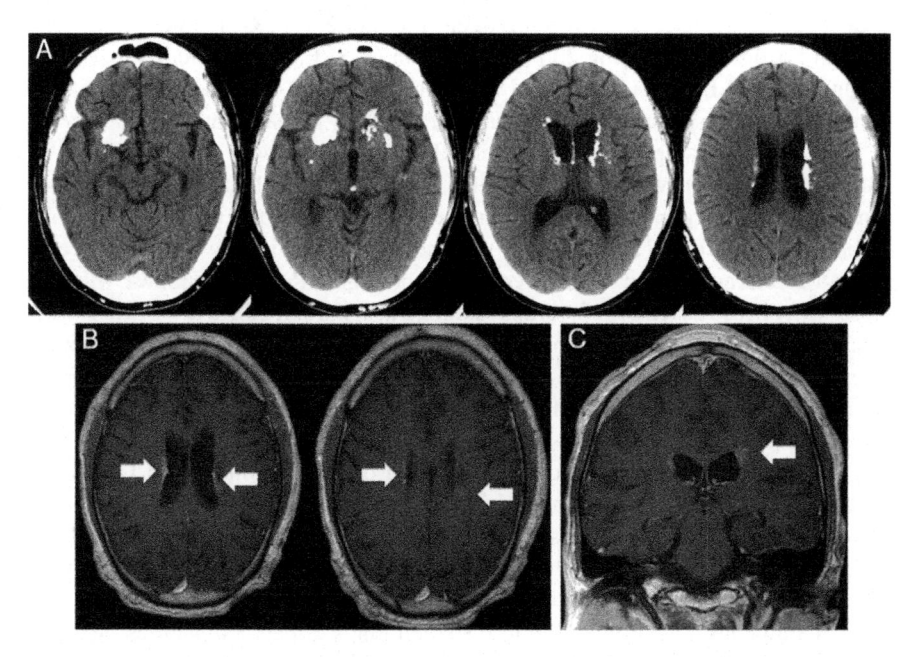

Fig. 8.2 Calcification due to vasculitis in a patient with SLE. Progressive calcification on basal ganglia and periventricular white matter on CT scan (**a**). The horizontal (**b**) and coronal sections (**c**) on MRI (T1-weghted) with Gd-enhancement, showing the leakage of contrast medium around periventricular area (arrows), suggesting that the presence of vasculitis

CT scan image is also advantageous to detect calcification. Brain calcification has been reported to be seen in 30% of NPSLE patients [6]. Most of these have basal ganglia calcifications, while white matter calcification is also seen. Shown is our patient who presented significant and progressive calcification in basal ganglia and periventricular white matter (Fig. 8.2). Previous studies have documented that inflammation including vasculitis may contribute to brain calcification [7, 8]. However, the precise mechanism of brain calcification in SLE remains unclear.

8.3 Magnetic Resonance Imaging (MRI)

MRI is the best available tool to detect abnormalities in the CNS including cerebrum, cerebellum, brain stem and spinal cord, because of its much greater resolution. In general, images obtained in MRI can be classified based on the examined conditions using each protocol including T1-weighted image, T2-weighted image, image with post gadolinium enhancement, fluid attenuated inversion recovery imaging (FLAIR), gradient recall echo (GRE) and diffusion weighted imaging (DWI), which are suggested to be useful for evaluation of NPSLE by the European League Against Rheumatism (EULAR) recommendation [2]. Sibbitt et al. documented clinico-pathological relationship between pre-mortem MRI finding and postmortem histopathology using 200 subjects with NPSLE, and concluded that brain lesions in NPSLE detected by MRI accurately represent serious underlying pre-mortem cerebrovascular and parenchymal brain injury on pathology [9].

Previous studies reported that brain MRI abnormalities were observed in 42.9–84.0% of the patients with NPSLE [10–15]. SLE patients with the antiphospholipid syndrome have the greater prevalence and severity of MRI abnormalities than those without this syndrome [16]. It is easy to remember that patients with NPSLE categorized as the neurologic syndromes should present MRI abnormalities, whereas those with diffuse psychiatric/neuropsychological syndrome (diffuse NPSLE) is not always complicated with abnormal finding on MRI. For example, cerebrovascular disease accompanied by antiphospholipid syndrome almost always has MRI abnormalities [17]. On the other hand, it has been revealed that 47.2% of patients with diffuse NPSLE had abnormal lesions revealed on brain MRI, even though they did not present with any focal neurological manifestations [18]. In this regard, Castellino et al. showed that abnormal brain MRI scans were observed in 55.8% of the patients with diffuse NPSLE, which was not significantly different from patients without NPSLE (36.6%) [19]. Thus, whether MRI abnormalities in diffuse NPSLE are non-specific or not needs further clarification.

MRI study has a great modality to reveal all kinds of parenchymal lesion especially in patients with neurologic syndromes of NPSLE, although availability is limited. A variety of detection methods such as T1-, T2-weighted, FLAR and DWI should be applied to ensure the characteristic of each lesion, which can help us to understand the activity and freshness of the lesion. Follow up study may also be important, because acute, reversible lesions could indicate the presence of active NPSLE even retrospectively. The characteristics of images in conventional MRI are highlighted in the following sections.

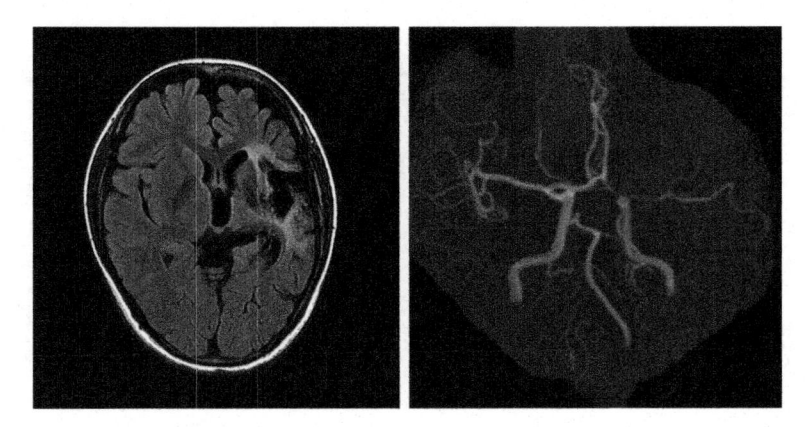

Fig. 8.3 Atrophy and infarction in a patient with SLE and antiphospholipid syndrome. FLAIR image presenting cortical atrophy around old infarcted area (left) and occlusion of left middle cerebral artery by MR angiography (right)

8.3.1 Parenchymal Lesion

Parenchymal lesions are representative MRI abnormalities presented over cerebral cortex and white mater. Small and multiple, disseminated and punctuated lesions in white matter are the most prevalent finding. DWI is very helpful to detect lesions at the early phase of infarcted area of the brain, and its abnormalities could be found within a few minutes after the onset when there is no finding in T2-weighted or FLAIR images [20–23]. As DWI abnormalities can occur in conditions other than infarction, apparent diffusion coefficient (ADC) mapping helps to confirm that DWI abnormalities is surely from ischemic lesion. ADC mapping is composed of the parameters obtained from DWI examined by different conditions. Therefore, ADC mapping should be referred to diagnose lesions when there is DWI abnormality.

The size of the infarction depends on the involved arterial size. Large size lesion is usually caused by thrombosis due to antiphospholipid syndrome (Fig. 8.1) [2]. Ischemia due to antiphospholipid syndrome might result in local atrophy of the brain (Fig. 8.3). Libman-Sacks endocarditis is a significant risk for stroke in SLE patients, developing embolic cerebrovascular disease [24]. Small size infarction may be due to occlusion of small artery, which could be resulted from thromboembolism as well as brain vasculitis, in which both new and old lesions caused by repeated infarctions are detected as small hyperintensity area in both DWI and T2-weighted images [9, 25–28].

Gadolinium (Gd)-enhancement is effective to demonstrate the presence of vasculitis, which can be useful as a follow-up study (Fig. 8.2). Vasculitis should be considered as a candidate which can develop ischemic abnormalities in brain other than cerebrovascular diseases caused by antiphospholipid antibodies in patients with SLE [27]. According to a study of primary CNS vasculitis, hyperintensity FLAIR lesion was observed in 95% patients with biopsy-proven vasculitis, whereas 67% patients had parenchymal lesions on images with Gd-enhancement [29].

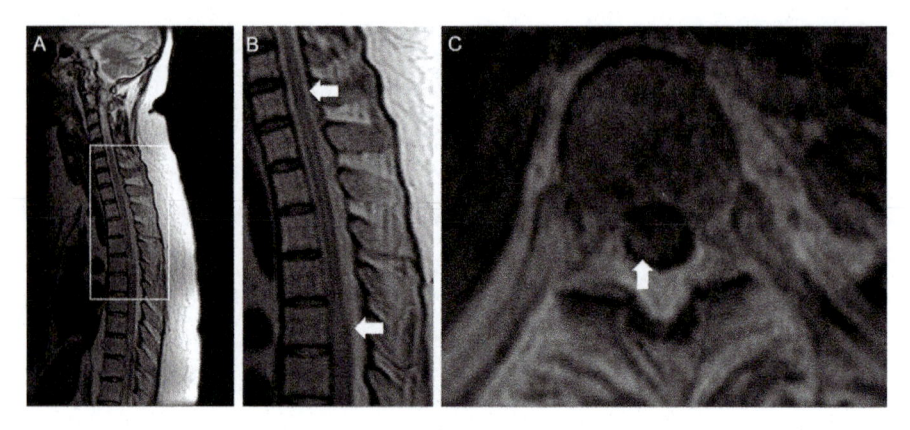

Fig. 8.4 NMOSD in an SLE patient. Longitudinal spinal cord lesion in a patient with SLE on T2-weighted MRI. The sagittal section showing abnormal hyperintensities from C6 to Th6 (**a**). High magnification of the sagittal section (**b**) and the horizontal section (**c**) showing a wide range of spinal hyperintensities (arrows) defined in the diagnostic criteria of NMOSD

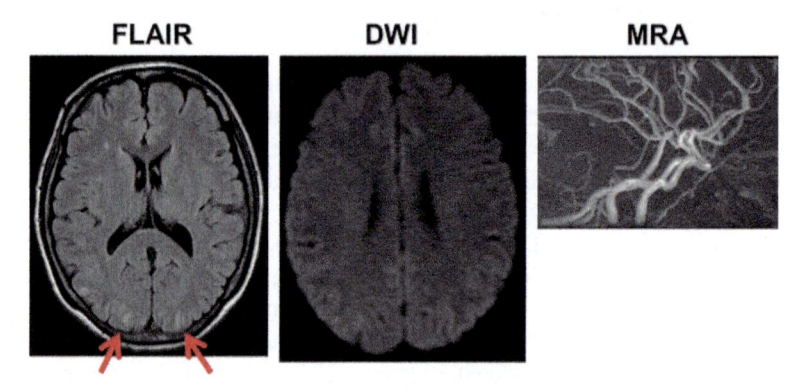

Fig. 8.5 PRES in SLE. Hyperintensity lesions in the bilateral posterior lobes in FLAIR images on MRI (Left) (arrows). No abnormalities in DWI images (Middle) or in MRA (Right)

Hemorrhagic lesion, especially intracranial hemorrhage, is also detected on MRI as hyperintensity lesion on T1-weighted images and as hypointensity lesion on T2-weighted images. In previous reports, cerebral hemorrhage in antiphospholipid syndrome was found to be associated with anti-platelet or anti-coagulation therapy [30], although the occurrence of cerebral hemorrhage has been shown to be related to antiphospholipid antibody itself [31].

As brainstem and spinal cord lesions involve a variety of clinical manifestations, MRI study is very important for making an accurate diagnosis. It should be noted that some patients who fit the criteria of demyelinating syndrome (Fig. 8.4) [32], might show positive serum anti-Aquaporin 4 antibodies, confirming the diagnosis of neuromyelitis optica spectrum disorders (NMOSD) [33]. A multicenter retrospective study has reported the association of NMOSD and NPSLE [34].

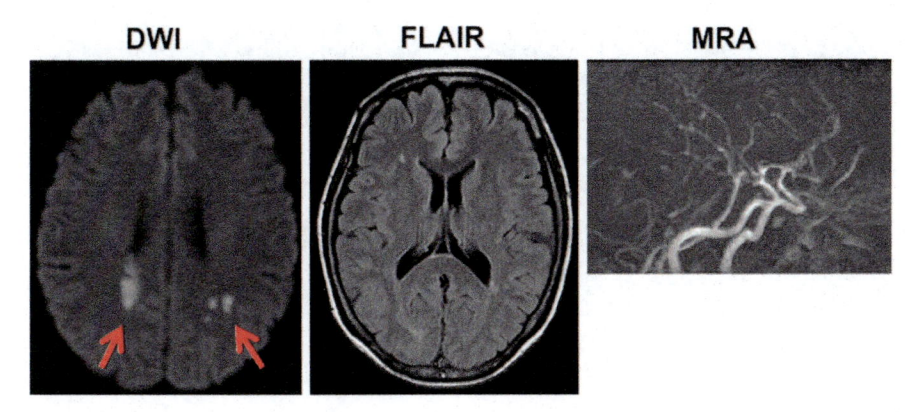

Fig. 8.6 RCVS in SLE. Hyperintensity lesions in the bilateral posterior white matters in DWI images on MRI (left) (arrows). Irregular narrowing of cerebral arteries in MRA, confirming the diagnosis of RCVS (right)

Posterior reversible encephalopathy syndrome (PRES) is occasionally observed in patients with SLE, for which younger age, history of seizure, hypertension and renal dysfunction are risk factors [35]. MRI scans show typical abnormalities including bilateral and asymmetrical isointensities or hypointensities on T1, hyperintensities in T2 and FLAIR sequences in the parietotemporo-occipital regions (Fig. 8.5) [36]. However, the precise relationship of PRES and NPSLE has not been explored.

Reversible cerebral vasoconstriction syndrome (RCVS) is developed by a reversible segmental and multifocal vasoconstriction of cerebral arteries with severe headaches with or without focal neurological deficits or seizures [37]. RCVS usually manifests 3 types of stroke, SAH, intracerebral hemorrhage, and cerebral infarction and also has reversible brain edema [37] with certain abnormalities on MRI (Fig. 8.6). As clinical features of RCVS resemble NPSLE, we need to consider RCVS as a differential diagnosis.

Apart from NPSLE, malignancy and infectious disease are crucial complications in SLE patients. Recently, CNS lymphoma case with SLE, especially associated with the use of mycophenolate mofetil has been reported [38–40], which can be obviously detected as an parenchymal lesion. For this reason, CNS lymphoma should be considered when MRI presents an unusual finding with an atypical clinical course for NPSLE (Fig. 8.7).

8.3.2 White Matter Hyperintensity (WMH)

WMH is the most typical finding on MRI in patients with NPSLE characterized by small, punctuated and usually multiple lesions ranging in size from 3 mm to 35 mm localized subcortical or periventricular white matter [10–12, 14] (Fig. 8.8). These lesions usually show hyperintensities on T2-weighted image and FLAIR, iso- or

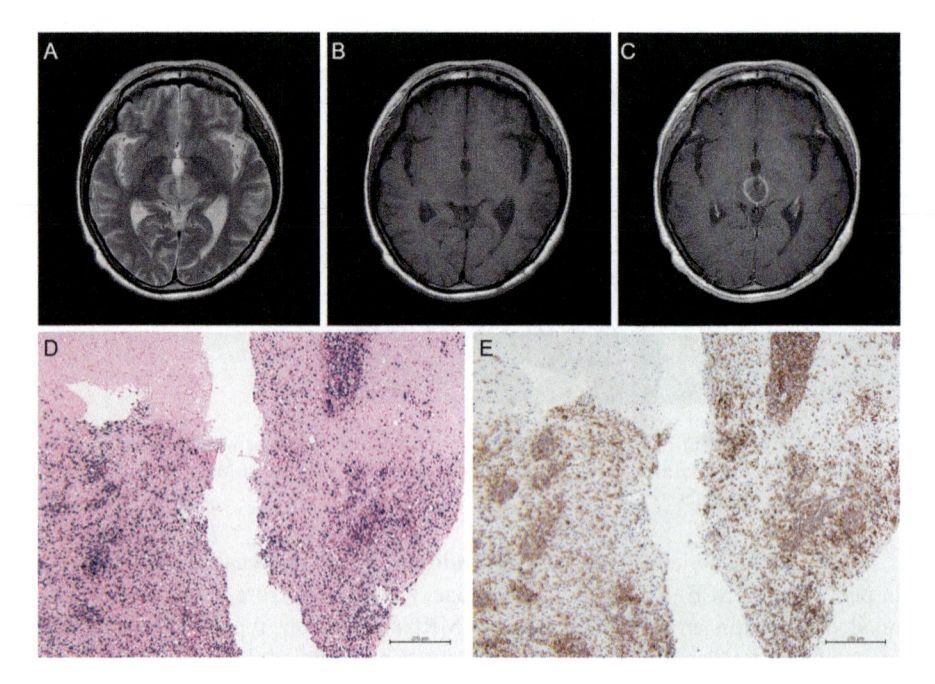

Fig. 8.7 CNS lymphoma in a lupus patient. Tumor-like lesion was observed as hyperintensity lesion on T2-weighted images (**a**) and as hypointensity lesion on T1-weighted image (**b**), which shows ring-enhancement by gadolinium (**c**). Biopsy revealed diffuse large B cell lymphoma. Hematoxylin-eosin stain (**d**) and immunohistostaining with anti-CD20 antibody (**e**)

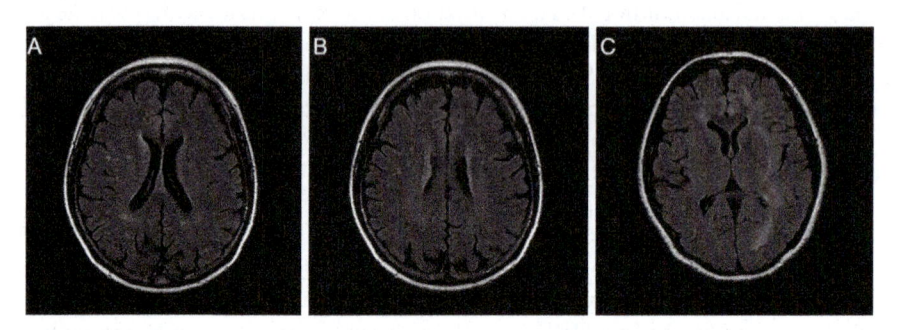

Fig. 8.8 White matter hyperintensity lesions in SLE. (**a**, **b**) Punctuated, multiple and small subcortical hyperintensities in white matter lesions are observed in an SLE patient, who also has bilateral periventricular hyperintensities on FLAIR images. (**c**) Extensive white matter hyperintensities seen in another SLE patient (the same patient as Fig. 8.2). These findings were ameliorated after immunosuppressive treatment, indicating the possible complication due to vasculitis

hypo intensities on T1-weighted image without high intensities on DWI, indicating that most lesions are chronic. FLAIR is more sensitive to detect WMH in patients with NPSLE, and the number of WMH detected by FLAIR were independently associated with NPSLE activity [41]. However, even in patients with diffuse NPSLE without obvious neurologic syndromes WMH was frequently observed (34.0%)

[18]. The prevalence of MRI abnormalities including WMH was similar between patients with diffuse NPSLE and SLE patients without NPSLE [19].

The presence of antiphospholipid antibodies was associated with an increased number of WMH [14]. According to quantitative analysis, headache, cognitive impairment and seizures were observed more frequently in SLE patients with WMH than those with normal MRI [14]. Therefore, in patients of SLE, punctuated WMH may be attributed to the disease activity or parts of NPSLE, although it could be non-specific. Indeed, abnormality in WMH is accelerated along with aging [42]. As another white matter lesion, diffuse infiltrative periventricular lesions are also common in NPSLE patients, observed in 6.1% of NPSLE patients and 3.8% of diffuse NPSLE [10, 18, 19] (Fig. 8.8).

8.3.3 Gray Matter Hyperintensity (GMH)

GMH is observed as larger diffuse hyperintensities (13–60 mm) and were located within the cortical gray matters [10], affecting the cortex and the basal ganglia [10, 18], although a few patients had small focal lesions of 3–11 mm in size within the cortex [10] (Fig. 8.9). Usually, GMH is observed with WMH and is detected T2-weightened image and FLAIR [10]. Generalized seizures could be associated with GMH [16, 43], accompanied by reversible focal and punctate high-intensity

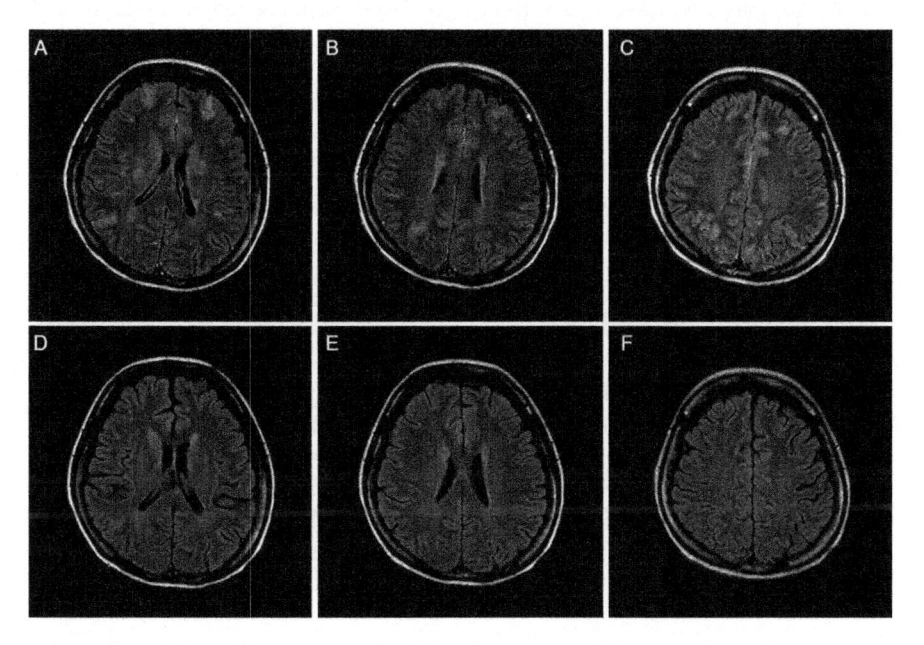

Fig. 8.9 MRI abnormalities in acute confusional state. (**a–c**) Multiple gray matter hyperintensities in cortex, along with white matter hyperintensities on FLAIR before treatment. (**d–f**) 14 days after administration of high dose corticosteroid. Abnormal hyperintensities were improved

lesions in both white and gray matters, which generally resolve rapidly and therefore must be studied quickly to be documented [16]. The mechanism of GMH is not clear, although a direct effect could be one of the causes developing GMH. Of note, even patients complicated with only diffuse NPSLE present GMH in basal ganglia and hippocampus [18]. Also, in some patients, GMH improved after immunosuppressive treatment. Thus, one of the causes developing GMH could be associated with a treatable lesion like vasculitis.

8.3.4 Atrophic Lesion

Atrophic lesion has been documented well in patients with SLE. The prevalence of atrophic lesion in patients with NPSLE is 5.4–50.0% with a variety of ranges [10–12, 19, 44]. Castellino et al. demonstrated that the prevalence of brain atrophy was not significantly different between NPSLE patients and SLE patients without NPSLE [19]. Moreover, Kozara et al. indicated that even SLE patients without previously obvious CNS manifestations had higher prevalence of brain atrophy than healthy individuals [45], although the other studies could not find such tendency [44, 46]. In patients with SLE, a reduction in the volume of cerebrum and corpus callosum is associated with disease duration, a history of CNS involvement, and cognitive impairment [44]. From the result of a population-based study in Finland, cerebral atrophy was associated with cognitive dysfunction, epileptic seizures, and cerebrovascular disease [47]. Of course, antiphospholipid syndrome or antiphospholipid antibodies affect the prevalence of brain atrophy through cerebrovascular hypoperfusion [16]. As localized findings on MRI, atrophy in hippocampus has been found, resulting in impairment of learning and memory [48–50]. However, in general, the association or the effect of brain atrophy with cognitive dysfunction in SLE patients is controversial [49, 50].

8.3.5 Meningeal Lesion

As a meningeal lesion on MRI, meningitis including hypertrophic pachymeningitis is the representative abnormality in patients with NPSLE, which can be detected by Gd-enhacement [2]. Recently, a few reports described hypertrophic pachymeningitis complicated in patients SLE as a rare complication [51, 52]. In aseptic meningitis, MRI with Gd-enhancement may help to find the portion of meningitis. However, since such abnormalities of enhanced-MRI are non-specific, it is necessary to discriminate other diseases.

8.4 Advanced Techniques of MRI

Although conventional MRI scan can give us meaningful information in CNS with high resolution images, these findings are not specific only for SLE or NPSLE. Recent new MRI techniques can provide the more specific results associated with NPSLE. These techniques include Magnetic resonance spectroscopy (MRS), diffusion tensor imaging (DTI), magnetization transfer imaging (MTI) and magnetic resonance relaxometry.

8.4.1 Magnetic Resonance Spectroscopy (MRS)

MRS is a technique of functional neuroimaging by evaluating a profile of neurochemical composition in brain. MRS data are usually displayed as spectra, with peaks reflecting the chemical structure and concentration of individual metabolites, including N-acetylaspatate (NAA), choline (Cho), myo-inositol (ml) and creatine [16]. Proton magnetic resonance spectroscopy is a common technique due to the requirement of additional scan only on conventional MRI [15]. However, NAA in NPSLE could be just associated with the degree of disseminated microinfarction seen on MRI, and is unanimously decreased in active and inactive NPSLE without improvement [15]. It is therefore suggested that the reduced NAA is reflecting loss of neurons, but not neuronal function disturbed by NPSLE [15]. A case report indicated ml elevation may predict the poor prognosis of NPSLE [53]. On the other hand, Cho peak elevation is indicated to be related to cognitive dysfunction in NPSLE even when there is no abnormality on conventional MRI, especially combined with the use of NAA [54–56]. Thus, MRS can be useful particularly to estimate the present damages of brain.

8.4.2 Diffusion Tensor Imaging (DTI)

DTI is an MRI technique based on detecting movement of water molecules. DTI is useful to find abnormality of white matter in brain and allows characterization of the microstructural properties and macroscopic organization of white matter tracts through measurement of fractional anisotropy (FA) and mean diffusivity (MD). FA is a marker for white matter fiber integrity and density, while MD is considered an indicator of brain maturation and/or injury [57]. Therefore, DTI seems to be useful to find white matter damage before appearance of abnormal finding on conventional MRI. Recently, a systematic review about the association of NPSLE and DTI concluded that the reduced FA values as well as the increased MD values, suggesting subclinical microstructural changes, are seen both in NPSLE and in non-NPSLE, compared with healthy control [58]. From this result, DTI could be advantageous to reveal subclinical damages in brain, although further studies are required to demonstrate the utility of DTI in NPSLE.

8.4.3 Magnetization Transfer Imaging (MTI)

MTI is superior to detect tissue damages, especially in white matter. By calculating magnetization transfer ratio (MTR), the amount of myelin could be measured, indicating that decreased MTR implicates demyelination. Many studies have demonstrated the usefulness of detecting active CNS damages in SLE patients by MTI [59–62]. Thus, Emmer et al. demonstrated that among various neuropsychiatric syndromes only cognitive dysfunction was associated with the MTR histogram peak height [61], whereas Magro-Checa et al. indicated that cognitive disorder, mood disorder, and psychosis were related to lower white matter MTR histogram peak heights values, while cerebrovascular symptoms were related to higher values [62]. Notably, MTI can be used to figure out the progression of brain damages as well as to evaluate the improvement of the damages by interval studies.

8.5 Single Photon Emission Computed Tomography (SPECT)

SPECT is a traditional method for evaluating brain function, using radio isotope, through detection of blood flow. SPECT used to be applied for evaluation of diffuse NPSLE. However, although its sensitivity is high enough [2], the specificity among patients with SLE is not so high to select NPSLE (sensitivity 70.3%, specificity 51.2%) [19]. Therefore, SPECT has now only a limited value as a neuroimaging study for NPSLE.

8.6 Positron Emission Tomography (PET)

PET is one of the tools in neuroimaging to evaluate brain glucose metabolism by injection of [(18)F]2-fluoro-2-deoxy-D-glucose (FDG). Recent studies have demonstrated the association of abnormalities in FDG-PET with a certain type of NPSLE. Thus, Saito et al. revealed that hypometabolism in the medial frontal gyrus may be related to major depressive disorder in SLE [63]. Moreover, the hippocampus and orbitofrontal cortex hypermetabolism detected by FDG-PET was correlated with impaired memory and mood alterations in SLE patients, compared to healthy controls [64]. However, since the availability of PET study is very limited, the data to evaluate the real efficacy of FDG-PET on NPSLE are too scarce.

8.7 Summary

Neuroimaging is very effective and potent to detect parenchymal and functional abnormalities in patients with NPSLE. CT is very useful for detecting an emergent lesion like acute hemorrhage within a few minutes. Conventional MRI plays an important role to look up a variety of CNS lesions with high resolution. WMH is common in patients with NPSLE, although the actual prevalence is not different among entire lupus patients. Advanced method with MRI can give us more information, especially on white matter lesion, earlier than conventional MRI. The interval examination by MRI after treatment of SLE is sometimes helpful to confirm the presence of vasculitis, especially when MRI abnormalities were ameliorated after treatment with steroids. The accumulation of data to evaluate the utility of FDG-PET is required in the future. It should be remembered that neuroimaging itself cannot confirm that the lesions are caused by NPSLE.

References

1. Ainiala H, et al. Validity of the new American College of Rheumatology criteria for neuropsychiatric lupus syndromes: a population-based evaluation. Arthritis Rheum. 2001;45:419–23.
2. Bertsias GK, et al. EULAR recommendations for the management of systemic lupus erythematosus with neuropsychiatric manifestations: report of a task force of the EULAR standing committee for clinical affairs. Ann Rheum Dis. 2010;69:2074–82.
3. Chang Y-S, et al. Increased risk of subarachnoid hemorrhage in patients with systemic lupus erythematosus: a nationwide population-based study. Arthritis Care Res (Hoboken). 2013;65:601–6.
4. Mimori A, et al. Subarachnoid hemorrhage and systemic lupus erythematosus. Lupus. 2000;9:521–6.
5. Liang MH, et al. The American College of Rheumatology nomenclature and case definitions for neuropsychiatric lupus syndromes. Arthritis Rheum. 1999;42:599–608.
6. Raymond AA, et al. Brain calcification in patients with cerebral lupus. Lupus. 1996;5:123–8.
7. Nordstrom DM, West SG, Andersen PA. Basal ganglia calcifications in central nervous system lupus erythematosus. Arthritis Rheum. 1985;28:1412–6.
8. Anderson JR. Intracerebral calcification in a case of systemic lupus erythematosus with neurological manifestations. Neuropathol Appl Neurobiol. 1981;7:161–6.
9. Sibbitt WL, et al. Magnetic resonance imaging and brain histopathology in neuropsychiatric systemic lupus erythematosus. Semin Arthritis Rheum. 2010;40:32–52.
10. Luyendijk J, et al. Neuropsychiatric systemic lupus erythematosus: lessons learned from magnetic resonance imaging. Arthritis Rheum. 2011;63:722–32.
11. Steup-Beekman GM, et al. Neuropsychiatric manifestations in patients with systemic lupus erythematosus: epidemiology and radiology pointing to an immune-mediated cause. Ann Rheum Dis. 2013;72(Suppl 2):ii76–9.
12. Karassa FB, et al. Predictors of clinical outcome and radiologic progression in patients with neuropsychiatric manifestations of systemic lupus erythematosus. Am J Med. 2000;109:628–34.
13. Sanna G, et al. Central nervous system involvement in systemic lupus erythematosus: cerebral imaging and serological profile in patients with and without overt neuropsychiatric manifestations. Lupus. 2000;9:573–83.
14. Appenzeller S, et al. Quantitative magnetic resonance imaging analyses and clinical significance of hyperintense white matter lesions in systemic lupus erythematosus patients. Ann Neurol. 2008;64:635–43.

15. McCune WJ, et al. Identification of brain lesions in neuropsychiatric systemic lupus erythematosus by magnetic resonance scanning. Arthritis Rheum. 1988;31:159–66.
16. Sibbitt WL, et al. Neuroimaging in neuropsychiatric systemic lupus erythematosus. Arthritis Rheum. 1999;42:2026–38.
17. Harris EN, et al. Cerebral disease in systemic lupus erythematosus. Springer Semin Immunopathol. 1985;8:251–66.
18. Arinuma Y, et al. Brain MRI in patients with diffuse psychiatric/neuropsychological syndromes in systemic lupus erythematosus. Lupus Sci Med. 2014;1:e000050.
19. Castellino G, et al. Single photon emission computed tomography and magnetic resonance imaging evaluation in SLE patients with and without neuropsychiatric involvement. Rheumatology (Oxford). 2008;47:319–23.
20. Walecki J, et al. MR in neurological syndromes of connective tissue diseases. Med Sci Monit. 2002;8:MT105–11.
21. Moritani T, et al. Diffusion-weighted echo-planar MR imaging of CNS involvement in systemic lupus erythematosus. Acad Radiol. 2001;8:741–53.
22. Bosma GPT, et al. Abnormal brain diffusivity in patients with neuropsychiatric systemic lupus erythematosus. AJNR Am J Neuroradiol. 2003;24:850–4.
23. Hjort N, et al. Ischemic injury detected by diffusion imaging 11 minutes after stroke. Ann Neurol. 2005;58:462–5.
24. Roldan CA, et al. Libman-sacks endocarditis and embolic cerebrovascular disease. JACC Cardiovasc Imaging. 2013;6:973–83.
25. Ellison D, et al. Intramural platelet deposition in cerebral vasculopathy of systemic lupus erythematosus. J Clin Pathol. 1993;46:37–40.
26. Brooks WM, et al. The histopathologic associates of neurometabolite abnormalities in fatal neuropsychiatric systemic lupus erythematosus. Arthritis Rheum. 2010;62:2055–63.
27. Arinuma Y, et al. Histopathological analysis of cerebral hemorrhage in systemic lupus erythematosus complicated with antiphospholipid syndrome. Mod Rheumatol. 2011;21:509–13.
28. Tono T, et al. Transverse myelitis extended to disseminated encephalitis in systemic lupus erythematosus: histological evidence for vasculitis. Mod Rheumatol. 2016;26:958–62.
29. De Boysson H, et al. Primary angiitis of the central nervous system: description of the first fifty-two adults enrolled in the french cohort of patients with primary vasculitis of the central nervous system. Arthritis Rheumatol. 2014;66:1315–26.
30. Boerma C, et al. An acute multiorgan thrombotic disorder associated with antiphospholipid antibodies; two "catastrophic" cases. Ann Rheum Dis. 1997;56(9):568.
31. Boura P, et al. Intracerebral hemorrhage in a patient with SLE and catastrophic antiphospholipid syndrome (CAPS): report of a case. Clin Rheumatol. 2005;24(4):420.
32. Magro Checa C, et al. Demyelinating disease in SLE: is it multiple sclerosis or lupus? Best Pract Res Clin Rheumatol. 2013;27:405–24.
33. Wingerchuk DM, et al. International consensus diagnostic criteria for neuromyelitis optica spectrum disorders. Neurology. 2015;85:177–89.
34. Mader S, et al. Understanding the antibody repertoire in neuropsychiatric systemic lupus erythematosus and neuromyelitis optica spectrum disorders: do they share common targets? Arthritis Rheumatol (Hoboken, NJ). 2017. doi:https://doi.org/10.1002/art.40356.
35. Merayo-Chalico J, et al. Clinical outcomes and risk factors for posterior reversible encephalopathy syndrome in systemic lupus erythematosus: a multicentric case-control study. J Neurol Neurosurg Psychiatry. 2016;87:287–94.
36. Hugonnet E, et al. Posterior reversible encephalopathy syndrome (PRES): features on CT and MR imaging. Diagn Interv Imaging. 2013;94:45–52.
37. Ducros A. Reversible cerebral vasoconstriction syndrome. Lancet Neurol. 2012;11:906–17.
38. Dasgupta N, et al. Central nervous system lymphoma associated with mycophenolate mofetil in lupus nephritis. Lupus. 2005;14:910–3.
39. Tsang HHL, et al. Diffuse large B-cell lymphoma of the central nervous system in mycophenolate mofetil-treated patients with systemic lupus erythematosus. Lupus. 2010;19(3):330.
40. Svobodova B, et al. Brain diffuse large B-cell lymphoma in a systemic lupus erythematosus patient treated with immunosuppressive agents including mycophenolate mofetil. Lupus. 2011;20:1452–4.

41. Sibbitt WL, et al. Fluid Attenuated Inversion Recovery (FLAIR) imaging in neuropsychiatric systemic lupus erythematosus. J Rheumatol. 2003;30:1983–9.
42. Raina A, et al. Cerebral white matter hyperintensities on MRI and acceleration of epigenetic aging: the atherosclerosis risk in communities study. Clin Epigenetics. 2017;9:21.
43. Jeong HW, et al. Brain MRI in neuropsychiatric lupus: associations with the 1999 ACR case definitions. Rheumatol Int. 2015;35:861–9.
44. Appenzeller S, et al. Cerebral and corpus callosum atrophy in systemic lupus erythematosus. Arthritis Rheum. 2005;52(9):2783.
45. Kozora E, et al. Magnetic resonance imaging abnormalities and cognitive deficits in systemic lupus erythematosus patients without overt central nervous system disease. Arthritis Rheum. 1998;41:41–7.
46. Appenzeller S, et al. Longitudinal analysis of gray and white matter loss in patients with systemic lupus erythematosus. NeuroImage. 2007;34:694–701.
47. Ainiala H, et al. Cerebral MRI abnormalities and their association with neuropsychiatric manifestations in SLE: a population-based study. Scand J Rheumatol. 2005;34:376–82.
48. Coan AC, et al. Quantification of hippocampal signal intensity in patients with mesial temporal lobe epilepsy. J Neuroimaging. 2003;13:228–33.
49. Appenzeller S, et al. Hippocampal atrophy in systemic lupus erythematosus. Ann Rheum Dis. 2006;65:1585–9.
50. Lapa AT, et al. Abnormality in hippocampal signal intensity predicts atrophy in patients with systemic lupus erythematosus. Lupus. 2017;26:633–9.
51. Sanchez-Garcia M, et al. Hypertrophic pachymeningitis associated with cerebral spinal fluid hypovolemia as initial presentation of systemic lupus erythematous. Lupus. 2014;23:197–200.
52. Han F, et al. Cranial and lumbosacral hypertrophic pachymeningitis associated with systemic lupus erythematosus: a case report. Medicine (Baltimore). 2016;95:e4737.
53. Guillen-Del Castillo A, et al. Increased myo-inositol in parietal white and gray matter as a biomarker of poor prognosis in neuropsychiatric lupus: a case report. Lupus. 2014;23:1073–8.
54. Huizinga TW, et al. Imaging modalities in central nervous system systemic lupus erythematosus. Curr Opin Rheumatol. 2001;13:383–8.
55. Sibbitt WL, et al. Neuropsychiatric systemic lupus erythematosus. Compr Ther. 1999;25:198–208.
56. Brooks WM, et al. Relationship between neurometabolite derangement and neurocognitive dysfunction in systemic lupus erythematosus. J Rheumatol. 1999;26:81–5.
57. Alexander AL, et al. Characterization of cerebral white matter properties using quantitative magnetic resonance imaging stains. Brain Connect. 2011;1:423–46.
58. Costallat BL, et al. Brain diffusion tensor MRI in systematic lupus erythematosus: a systematic review. Autoimmun Rev. 2018;17:36–43.
59. Bosma GP, et al. Evidence of central nervous system damage in patients with neuropsychiatric systemic lupus erythematosus, demonstrated by magnetization transfer imaging. Arthritis Rheum. 2000;43:48–54.
60. Bosma GP, et al. Detection of cerebral involvement in patients with active neuropsychiatric systemic lupus erythematosus by the use of volumetric magnetization transfer imaging. Arthritis Rheum. 2000;43:2428–36.
61. Emmer BJ, et al. Correlation of magnetization transfer ratio histogram parameters with neuropsychiatric systemic lupus erythematosus criteria and proton magnetic resonance spectroscopy: association of magnetization transfer ratio peak height with neuronal and cognitive dy. Arthritis Rheum. 2008;58:1451–7.
62. Magro-Checa C, et al. Changes in white matter microstructure suggest an inflammatory origin of neuropsychiatric systemic lupus erythematosus. Arthritis Rheumatol (Hoboken, NJ). 2016;68:1945–54.
63. Saito T, et al. Regional cerebral glucose metabolism in systemic lupus erythematosus patients with major depressive disorder. J Neurol Sci. 2017;379:127–30.
64. Mackay M, et al. Brain metabolism and autoantibody titres predict functional impairment in systemic lupus erythematosus. Lupus Sci Med. 2015;2:e000074.

Chapter 9
Psychiatric Symptoms

Katsuji Nishimura

Abstract Psychiatric symptoms often occur in patients with systemic lupus erythematosus (SLE). According to the American College of Rheumatology (ACR) research committee criteria for neuropsychiatric SLE (NPSLE) in 1999, the psychiatric symptoms include acute confusional state (delirium), anxiety disorder, cognitive dysfunction, mood disorder, and psychosis. Because no diagnostic gold standard exists for primary NPSLE, it is often difficult to distinguish primary cause-and-effect association of the symptom as a direct attribute of active disease from a secondary, indirect effect resulting from complications of the disease or its therapy (e.g. corticosteroids), or an effect unrelated to SLE. On one hand, conditions such as psychosis, anxiety disorder and mood disorder can also present as psychological reaction to the disease and the related stress. On the other hand, possible mechanisms include microvasculopathy and thrombosis, or autoantibodies and inflammatory mediators. Therefore, in lupus patients presenting with psychiatric symptoms, a careful diagnostic work-up including, but not limited to, psychiatric/neuropsychological assessment, immunological/medical investigation, as well as neuroimaging is crucial in delivering adequate medical care and treatment.

Keywords Lupus Psychosis · Depression · Anxiety · Acute confusional state · Cognitive dysfunction

---

K. Nishimura (✉)
Department of Psychiatry, Tokyo Women's Medical University School of Medicine, Tokyo, Japan
e-mail: nishimura.katsuji@twmu.ac.jp

9.1 Introduction

Systemic lupus erythematosus (SLE), a chronic and relapsing-remitting autoimmune inflammatory disease, affects multiple physiological systems including the central nervous system (CNS). A variety of neurologic and psychiatric symptoms frequently occur in patients with SLE [1–4]. In 1999, the American College of Rheumatology (ACR) research committee published case definitions for 19 neuropsychiatric SLE (NPSLE) symptoms consisting of 12 CNS- related and 7 peripheral nervous system (PNS) -related symptoms [5]. Five of the 12 CNS-related symptoms were psychiatric symptoms that included acute confusional state, anxiety disorder, cognitive dysfunction, mood disorder, and psychosis. These psychiatric symptoms are associated with a negative impact on prognosis, quality of life, overall damage of the disease, and working disability [6].

9.2 Primary and Secondary NPSLE

Despite a high incidence of psychiatric manifestations associated with SLE, it is unclear whether this association is a direct consequence of microvasculopathy and thrombosis, or of autoantibodies and inflammatory mediators (primary NPSLE). Equally unclear is whether the association reflects an indirect effect (secondary NPSLE) resulting from complications of the disease (e.g. uremia or hypertension) or therapy (e.g. corticosteroids), or an effect unrelated to SLE (e.g. infections, metabolic abnormalities, or adverse medication effects) [7, 8]. The 1999 ACR classification of NPSLE [5] was not necessarily specific for primary NPSLE because no diagnostic gold standard of primary NPSLE existed [8]. To provide further insight in the understanding of primary and secondary NPSLE, Unterman et al. [4] conducted a meta-analysis of 17 studies on NPSLE applying the 1999 ACR case definitions, and estimated the prevalence of the psychiatric symptoms as follows: mood disorder (20.7%), cognitive dysfunction (19.7%), anxiety disorder (6.4%), psychosis (4.6%), and acute confusional state (3.4%) (Table 9.1).

In order to assess the validity of the ACR nomenclature for NPSLE, Ainiala et al. [1] conducted a cross-sectional, population-based study covering an area with 440,000 people. A total of 46 SLE patients and 46 controls matched by age, sex, education, and place of residence underwent a clinical neurologic examination and neuropsychological testing. Forty-two (91%) of 46 patients and 25 (56%) of 46 controls fulfilled at least one of the ACR NPSLE criteria. Cognitive dysfunction was the most common syndrome detected in 37 patients (80%). They observed a myriad of NP events that also occurred with high frequency in normal population controls: anxiety, mild depression (that failed to meet the criteria for "major depressive-like episodes"), and mild cognitive impairment (deficits in fewer than three of the eight specified cognitive domains).

Table 9.1 Estimated prevalence of psychiatric symptoms in SLE

Psychiatric symptoms[a]	Estimated prevalence[b]	
	%	95% CI
Mood disorder	20.7	11.5–37.4
Cognitive dysfunction	19.7	10.7–36.0
Anxiety disorder	6.4	3.0–13.6
Psychosis	4.6	2.4–8.8
Acute confusional state	3.4	1.1–10.3

SLE systemic lupus erythematosus, *CI* confidence interval
[a] Defined by the American College of Rheumatology nomenclature and case definitions for neuropsychiatric lupus syndromes (1999)
[b] Estimated by meta-analysis using random-effects model by Unterman et al. [4]

Table 9.2 Psychiatric symptoms and the ratio of the attribution to SLE

Psychiatric symptoms[a]	No. of primary NPSLE events/No. of total NP events (%)[b]			
	Most stringent NPSLE		Least stringent NPSLE	
Mood disorder	18/139	(12.9%)	47/139	(33.8%)
Cognitive dysfunction	8/43	(18.6%)	22/43	(51.2%)
Anxiety disorder	0/42	(0%)	0/42	(0%)
Psychosis	8/14	(57.1%)	13/14	(92.9%)
Acute confusional state	11/22	(50%)	17/22	(77.3%)

NPSLE neuropsychiatric systemic lupus erythematosus
[a] Defined by the American College of Rheumatology nomenclature and case definitions for neuropsychiatric lupus syndromes (1999)
[b] From an international disease inception cohort of SLE patients based on the attribution model by Hanly et al. [11]

In this context, Hanly et al. [9] developed an attribution model using rules of different stringency based on (1) the onset of NP events prior to the diagnosis of SLE, (2) concurrent non-SLE factors identified from the ACR glossary for each NP syndrome, and (3) exclusion of "common" NP events (anxiety, mild depression, and mild cognitive impairment) as described above by Ainiala et al. [1].

In a prospective, single-center cohort study of up to 7 years, 132 (63%) of 209 patients were demonstrated to show at least one NP event (299 events total) characterized by the ACR case definitions, but only 31% of the total events were attributed to SLE [10]. Furthermore, an international inception cohort study showed that 486 (40.3%) of 1206 patients had at least one NP event over a mean follow-up period of 1.9 ± 1.2 years [11]. Also, the proportion of NP events attributed to SLE varied from only 13.0% (the most stringent model of NPSLE) to 23.6% (the least stringent model of NPSLE) of patients. The psychiatric events most frequently attributed to SLE include psychosis (57.1% [the most stringent] to 92.9% [the least stringent]), acute confusional states (50% to 77.3%) and cognitive dysfunction (18.6% to 51.2%). The psychiatric events least frequently attributed to SLE were anxiety disorders (0%) and mood disorders (12.9% to 33.8%) (Table 9.2).

9.3 SLE-Associated Psychiatric Syndromes

Five psychiatric symptoms were included in the 1999 ACR case definitions of NPSLE that included psychosis, mood disorder, anxiety disorder, acute confusional state, and cognitive dysfunction. Psychosis, mood- and anxiety- disorders were collectively called psychiatric disorders [5]. In addition, other psychiatric disorders or conditions in connection with NPSLE, e.g., catatonia or suicide behaviors have been also reported in the literature.

9.3.1 Psychiatric Disorders

Lupus patients frequently experience a variety of psychiatric disorders including psychosis, mood- and anxiety- disorders. As mentioned above, the psychobiological relationship between these psychiatric disorders and SLE is actually poorly understood. These disorders, therefore, are problematic from the perspective of diagnostic workup and etiology, particularly with regards to whether they are manifestation(s) of primary or secondary NPSLE.

The ACR case definition of these psychiatric disorders [5] was based on the Diagnostic and Statistical Manual of Mental Disorders, 4th edition (DSM-IV) entity as "psychosis/mood disorder/anxiety disorder due to a general medical condition" [12]. There is some support for the idea that the disturbance that manifests as NPSLE is not better accounted for by another mental disorder (e.g. psychological reaction to the stress of having SLE), does not occur exclusively during the course of delirium, and is severe enough to cause significant distress or social impairment [12]. Hence, in order to satisfy the criteria of DSM-IV entity, there needs evidence from the history, physical examination, or laboratory findings that the disturbance may be the direct physiological consequence of the medical condition. However, demonstrating such evidence is often difficult in SLE patients clinically presenting with psychiatric disorders.

9.3.1.1 Psychosis

The essential feature of psychosis is prominent hallucinations or delusions [5, 12]. In a single-center cohort study from UK [13], psychosis due to lupus was diagnosed in 11 (2.3%) of 485 patients with SLE. Psychosis presented as the initial manifestation of SLE in 60%, and within the first year of the disease in 80% of the cases. All the patients developed psychosis within the context of multiple systemic lupus activity and after intensive immunosuppressive treatment, 70% of the cases showed complete resolution of the psychosis (although chronic mild psychotic symptoms were observed in 30% of the cases).

A retrospective investigation showed that acute psychosis occurred in 89 (17%) out of a cohort of 520 patients with SLE from the view point of differential diagnosis of corticosteroid-induced psychosis [14]. Also, psychosis primary to CNS involvement was diagnosed in 59, corticosteroid-induced psychosis in 28, and primary psychotic disorder unrelated to SLE or medication in 2 patients. In addition, psychosis due to lupus at the onset of SLE was observed in 19 patients and was directly associated with lupus disease activity. Furthermore, the psychosis during follow-up of SLE was observed in 40 patients and associated with positive antiphospholipid antibodies and less frequently with renal and cutaneous involvement.

In a recent retrospective study from Thailand [15], 36 episodes of psychosis or psychotic depression were identified in 35 (5%) of 750 patients with SLE. Eleven episodes (31%) occurred during the first manifestation of lupus. The psychotic symptoms included persecutory delusion (50% of the episodes), bizarre delusion (44%), third person auditory hallucinations (44%) and visual hallucinations (36%). Twenty-four episodes (67%) were associated with active lupus in CNS and other organs. In one case, death resulted from suicide although most of the psychotic episodes (97%) showed complete remission with rare recurrences. Depressive psychosis required psychotropic treatment longer than psychosis alone.

Anti-ribosomal P antibody has been demonstrated to be associated with lupus psychosis related to SLE disease activity [16, 17], but its clinical usefulness in the diagnosis of lupus psychosis has been impeded by its low sensitivity [18]. On the contrary, the high accuracy of cerebrospinal fluid interleukin-6 testing for diagnosis of lupus psychosis was demonstrated in a Japanese multi-center retrospective study [19]. In this study, the sensitivity and specificity for diagnosis of lupus psychosis were 88% and 92%, respectively, at the cut-off value of 4.3 pg/ml, hence making the CSF interleukin-6 level a potent functional and clinical index for the lupus diagnosis and treatment.

Taken together, while lupus psychosis is not common, it usually occurs early in the course of the disease and is often associated with SLE direct disease activity, although its immunological surrogate markers are limited. Hence, lupus psychosis usually can be improved by intensive immunosuppressive treatment in addition to antipsychotics [13–15].

9.3.1.2 Mood Disorder

The essential feature of mood disorders is prominent and persistent disturbance in mood and is characterized by either (or both) of the following: (1) depressive features (depressed mood or markedly diminished interest or pleasure in all, or almost all, activities); (2) manic features (elevated, expansive, or irritable mood) [5, 12].

A recent meta-analysis [20] revealed that the prevalence of major depression in SLE patients was 24% (95% CI, 16%–31%) according to clinical interviews. The prevalence estimates of depression were 30% (95% CI, 22%–38%) on the Hospital Anxiety and Depression Scale with thresholds of 8 and 39% (95% CI, 29%–49%) for the 21-Item Beck Depression Inventory with thresholds of 14, respectively.

A systemic review about depression in SLE [21] demonstrated that the depressive symptoms frequently reported were fatigue and weakness (88–90%), irritability (82%), somatic preoccupation (76%) and trouble falling asleep (70%). Also, sadness has been reported from 29% to 77% of patients assessed.

According to a recent review [21], the most frequent possible cause of depression in SLE has been considered to be psychosocial factors including psychological reaction to the disease and social stress. On the other hand, some studies have suggested a relationship between depression and disease activity of SLE [22, 23]. Several studies suggested a possible involvement of autoantibodies, including antiribosomal P antibodies [23, 24], antineuronal antibodies and antiphospholipid antibodies [22], in the pathogenesis of depression.

A few neuroimaging studies in depressed lupus patients have also been conducted. For example, a study with single photon emission computed tomography (SPECT) [25] demonstrated that depressed patients with SLE have cerebral blood flow reductions in discrete temporal and frontal regions that may account for depressive symptoms. Similarly, another recent study using positron emission tomography (PET) [26] revealed hypometabolism in the medial frontal gyrus that may be related to major depressive disorder in SLE patients.

It has been deduced, based on evidence, that a large majority of the mood disorders of NPSLE are depression while mania is rarely present, thus occurring in approximately only 3% of patients [2].

9.3.1.3 Anxiety Disorder

The essential feature of anxiety disorders includes a prominent anxiety, panic attacks, and obsessive- compulsive disorders in the clinical perspective [5, 12]. A meta-analysis [20] demonstrated that the prevalence of anxiety was 37% (95% CI, 12%–63%) according to clinical interviews. The corresponding pooled prevalence was 40% (95% CI, 30%–49%) for anxiety according to the Hospital Anxiety and Depression Scale with a cutoff score of 8 or more. One study including 326 women with lupus, using a self-reported questionnaire [22] demonstrated that the lifetime prevalence of specific phobia (24%), panic disorder (16%), and obsessive-compulsive disorder (9%) were more common among patients with SLE than among other white women. Consequently, this study suggested the existence of a relationship between anxiety disorders and disease activity of SLE, although such a relationship has not been confirmed in most other studies.

9.3.2 Acute Confusional State (Delirium)

The acute confusional state is equivalent to "delirium," as defined in DSM-IV [12], referring to a disturbance of consciousness or level of arousal with reduced ability to focus, maintain, or shift attention, accompanied by cognitive disturbance and/or changes in mood, behavior, or affect [5]. The symptom usually develops over a short period of time, tends to fluctuate during the course of the day, and encompasses a spectrum from mild disturbances of consciousness to coma or unconsciousness [5]. The term "encephalopathy" or "acute organic brain syndrome" has also been used to describe the same clinical state.

Our group [27] revealed that interleukin-6 levels in the CSF as well as the IgG index showed statistically significant associations with acute confusional state in SLE patients, although no other single CSF test had sufficient predictive value to diagnose acute confusional state in SLE. Thus, the use of CSF tests combined with careful history and clinical examinations is recommended for proper diagnosis of acute confusional state in SLE.

9.3.3 Cognitive Dysfunction

In lupus patients, mild or moderate cognitive dysfunction is common, while severe dysfunction is relatively uncommon [7]. For example, one study investigating episodic verbal learning and memory using the Hopkins Verbal Learning Test-Revised, in 741 patients over 65 years of age with SLE, demonstrated that 452 (61%) patients had intact memory function (−0.5 standard deviations [SD] or higher than the age-matched normative value), 202 (27%) had mild to moderate impairment (−0.5 to −2.0 SD), and 87 (12%) had severe impairment (less than −2.0 SD or below age-matched population means) [28].

Cognitive dysfunction is confirmed by formal neuropsychological testing batteries that examine various domains of cognitive functions. In the 1999 ACR case definitions, these domains included simple attention, complex attention, memory, visual-spatial processing, language, reasoning and/or problem solving, psychomotor speed, and executive functions (such as multitasking, organization or planning). The committee recommended a short battery, lasting approximately 1 h for the assessment of these conditions. To satisfy the ACR case definition of cognitive dysfunction, at least one of these eight domains must be affected [5].

The Ad Hoc Committee of the ACR [29] conducted a systematic review of 25 controlled studies using the ACR neurocognitive short battery. It was demonstrated that significantly lower levels were found only in 3 of the 8 domains, namely attention, memory, and psychomotor speed, in SLE patients when compared with controls. The committee proposed that "cognitive impairment" was a deficit of 2.0 or more SD below the mean, compared with normative data, in these 3 key domains, and "cognitive decline" was consequently defined as a deficit of 1.5–1.9 SD below the mean.

The SLE-associated cognitive dysfunction may be a residual factor in patients with previous CNS impairments, or may serve as an early marker of CNS impairments in patients without NP symptoms. This proposition is suggested by the fact that cognitive dysfunction has been reported both in SLE patients with (40–60%) and without (20–30%) overt NP symptoms [30]. In addition, SLE-associated cognitive dysfunction may result from several conditions other than SLE, including psychiatric disturbances, pain, fatigue, sleep disturbance, or medications. For example, it has been shown that corticosteroids may affect the incidence and profile of cognitive dysfunction in patients with SLE, linking corticosteroids to cognitive dysfunction, especially verbal memory [31].

Results from our recent prospective study [32] suggested that psychomotor slowing may be a primary characteristic of cognitive dysfunction in lupus patients. In that study, we investigated cognitive dysfunction in corticosteroid-naïve patients with active, early-stage SLE. We found that the dominant characteristic of the cognitive dysfunction was slower psychomotor speed (assessed using the Digit Symbol Substitution Test), that was associated with increased SLE disease activity. Other impairments such as verbal memory deficits were not evident among such patients [32].

Of note, the psychomotor clinical presentation may reflect white matter inflammation in early SLE, because a prominence of cerebral white matter abnormalities as a general feature of cognitive slowing has been demonstrated [33]. In addition, observations in recent neuroimaging studies, such as magnetic resonance spectroscopy, showed inflammatory changes in white matter even in SLE patients who had no overt NP manifestations [34, 35], or who were newly diagnosed as SLE with no focal neurologic syndromes [36]. These data underscore the role of microstructural white matter changes in the cognitive impairment of non-NPSLE patients. Notably, persistently positive antiphospholipid antibodies have been reported as risk factors for cognitive dysfunction in lupus patients, especially for moderate to severe cases [37, 38].

9.3.4 Other Psychiatric Symptoms and Conditions

9.3.4.1 Catatonia

Catatonia is a rare manifestation of NPSLE. It can be classified into psychosis among 5 syndromes of diffuse NPSLE. A review of relevant literature [39] showed the most consistently reported symptoms were mutism, posturing, withdrawal, negativism, hypertonia and staring, while the most common psychiatric morbidity associated with catatonia has been affective disorders. Catatonia can present as initial symptoms of SLE or as a relapse of the disease. For instance, the treatment of lupus catatonia with electroconvulsive therapy [40] and benzodiazepines [41] has been reported as a useful option besides the use of intensive immunosuppressive therapy.

9.3.4.2 Suicide Ideation and Attempts

The ideation and incidence of suicide have been reported as ranging from 8% to 25% in lupus patients [42–44]. Mok et al. [43] revealed that the risk factors of suicidal thoughts in 367 lupus patients were depressive symptoms, cardiovascular damage, recent life events and previous suicide attempts. In another study, depression, anxiety, and patients' subjective complaints have been demonstrated as risk factors for suicidal ideation [44].

Suicide attempts occur more often in lupus patients compared with the general population. For example, a national-wide study from Taiwan [45] demonstrated that the incidence of drug overdose as suicide attempts in SLE patients and in general population were 291 and 160 cases per 100,000 person-years, respectively. The suicidal drug overdose was associated with psychiatric disorders such as depressive disorders and insomnia (mood disorder in the ACR criteria), and lower monthly income.

9.4 Special Consideration of Corticosteroid-Induced Psychiatric Disorder

It is known that corticosteroids, cornerstone of the treatment SLE, often induce a variety of psychiatric symptoms including manic/depressive mood changes, psychosis, delirium, anxiety, or impaired memory. Hence, the differential diagnosis of NPSLE and corticosteroid-induced psychiatric disorder (CIPD) remains challenging in clinical practice [46].

Two prospective cohort studies [47, 48] employed the following general strict definition of CIPD in patients with SLE: new-onset of psychiatric symptoms that developed within 8 weeks of initiation or augmentation of corticosteroid therapy and that resolved completely through a reduction in corticosteroid dosage and without additional immunosuppressive agents. They demonstrated that the incidence of CIPD in SLE patients ranged from 5% [47] to 10% [48]. In one study [47], 3 of the 6 cases with CIPD developed psychosis and the other 3 developed mood disorders with manic features. In the other study [48], mood disorders occurred in 13 of the total 14 cases with CIPD, including dominantly manic features in 9 cases. The presentation of psychosis occurred only in one patient.

Regarding the risk factors for CIPD in lupus patients, hypoalbuminemia has been reported [14, 47]. Interestingly, our group [48] demonstrated positive Q-albumin (cerebrospinal fluid/serum albumin ratio; an indicator of blood-brain barrier damage) as an independent risk factor for CIPD in SLE patients. However, it should be noted that an even higher level of Q albumin was noted in episodes with active diffuse NPSLE. Therefore, it is not possible to discriminate CIPD from diffuse NPSLE by Q albumin. It appears that CSF interleukin-6 might be a good tool for a differential diagnosis between CIPD and diffuse NPSLE [19].

The situation is also complicated because of the empirically well-known fact that new-onset psychiatric symptoms in lupus patients on corticosteroids could be caused by NPSLE and not by a CIPD. In the observational study that we conducted [49], during the 8 weeks of corticosteroid administration, new psychiatric events occurred in 20 (14.4%) of the 139 episodes in 135 patients with a non-NPSLE flare. Of the 20 patients, 2 (both presenting delirium) were diagnosed as CNS-SLE on the basis of evidence of abnormal CNS findings (e.g., EEG diffuse slowing) even before psychiatric manifestations, all of which improved in parallel with these patients' recoveries through augmentation of immunosuppressive therapy. This study suggests that corticosteroid therapy triggers CIPD and NPSLE in patients with SLE at the same time. Therefore, these 2 conditions may not be necessarily antinomic.

Psychopharmacotherapy may be required depending on the severity of CIPD, particularly if dose reduction or discontinuation of corticosteroids is impossible. However, the evidence has been very limited. For both psychotic and manic/mixed episodes of CIPD, several mood stabilizers (e.g., lithium or valproate) and antipsychotics (e.g., haloperidol or risperidone) have been effectively used in several case reports with good tolerance [50].

9.5 Summary

Psychiatric complications are common in patients with SLE. Cognitive dysfunction and depression (mood disorder) appear to be the most common manifestations. Acute confusional state (delirium), anxiety disorder, and psychosis also occur in lupus patients. To date, the nature of psychobiological relationship between these psychiatric symptoms and SLE is still unclear, although the roles of several autoantibodies have been implicated. In future studies, appropriately controlled studies are needed to clarify the pathogenetic mechanism of SLE-associated psychiatric symptoms and to help guide a decision-making for the treatment and prevention.

References

1. Ainiala H, et al. The prevalence of neuropsychiatric syndromes in systemic lupus erythematosus. Neurology. 2001;57:496–500.
2. Brey RL, et al. Neuropsychiatric syndromes in lupus: prevalence using standardized definitions. Neurology. 2002;58:1214–20.
3. Hanly JG, et al. Neuropsychiatric events in systemic lupus erythematosus: attribution and clinical significance. J Rheumatol. 2004;31:2156–62.
4. Unterman A, et al. Neuropsychiatric syndromes in systemic lupus erythematosus: a meta-analysis. Semin Arthritis Rheum. 2011;41:1–11.
5. ACR Ad Hoc Committee on Neuropsychiatric Lupus Nomenclature. The American College of Rheumatology Nomenclature and Case Definitions for neuropsychiatric lupus syndromes. Arthritis Rheum. 1999;42:599–608.
6. Postal M, et al. Neuropsychiatric manifestations in systemic lupus erythematosus: epidemiology, pathophysiology and management. CNS Drugs. 2011;25:721–36.
7. Bertsias GK, Boumpas DT. Pathogenesis, diagnosis and management of neuropsychiatric SLE manifestations. Nat Rev Rheumatol. 2010;6:358–67.
8. Hanly JG. Diagnosis and management of neuropsychiatric SLE. Nat Rev Rheumatol. 2014;10:338–47.
9. Hanly JG, et al. Short-term outcome of neuropsychiatric events in systemic lupus erythematosus upon enrollment into an international inception cohort study. Arthritis Rheum. 2008;59:721–9.

10. Hanly JG, et al. Prospective study of neuropsychiatric events in systemic lupus erythematosus. J Rheumatol. 2009;36:1449–59.
11. Hanly JG, et al. Systemic Lupus International Collaborating Clinics (SLICC): prospective analysis of neuropsychiatric events in an international disease inception cohort of patients with systemic lupus erythematosus. Ann Rheum Dis. 2010;69:529–35.
12. American Psychiatric Association, editor. Diagnostic and statistical manual of mental disorders. 4th ed. Washington, DC: American Psychiatric Association; 1994.
13. Pego-Reigosa JM, Isenberg DA. Psychosis due to systemic lupus erythematosus: characteristics and long-term outcome of this rare manifestation of the disease. Rheumatology. 2008;47:1498–502.
14. Appenzeller S, et al. Acute psychosis in systemic lupus erythematosus. Rheumatol Int. 2008;28:237–43.
15. Paholpak P, et al. Characteristics, treatments and outcome of psychosis in Thai SLE patients. J Psychosom Res. 2012;73:448–51.
16. Isshi K, Hirohata S. Differential roles of the anti-ribosomal P antibody and antineuronal antibody in the pathogenesis of central nervous system involvement in systemic lupus erythematosus. Arthritis Rheum. 1998;41:1819–27.
17. Toubi E, Shoenfeld Y. Clinical and biological aspects of anti-P-ribosomal protein autoantibodies. Autoimmun Rev. 2007;6:119–25.
18. Karassa FB, et al. Accuracy of anti-ribosomal P protein antibody testing for the diagnosis of neuropsychiatric systemic lupus erythematosus: an international meta-analysis. Arthritis Rheum. 2006;54:312–24.
19. Hirohata S, et al. Accuracy of cerebrospinal fluid IL-6 testing for diagnosis of lupus psychosis. A multicenter retrospective study. Clin Rheumatol. 2009;28:1319–23.
20. Zhang L, et al. Prevalence of depression and anxiety in systemic lupus erythematosus: a systematic review and meta-analysis. BMC Psychiatry. 2017;17:70.
21. Palagini L, et al. Depression and systemic lupus erythematosus: a systematic review. Lupus. 2013;22:409–16.
22. Bachen LA, et al. Prevalence of mood and anxiety disorders in women with systemic lupus erythematosus. Arthritis Rheum. 2009;61:822–9.
23. Nery FG, et al. Prevalence of depressive and anxiety disorders in systemic lupus erythematosus and their association with antiribosomal P antibodies. Prog Neuro-Psychopharmacol Biol Psychiatry. 2008;32:695–700.
24. Eber T, et al. Anti-ribosomal P-protein and its role in psychiatric manifestations of systemic lupus erythematosus: myth or reality? Lupus. 2005;14:571–5.
25. Giovacchini G, et al. Cerebral blood flow in depressed patients with systemic lupus erythematosus. J Rheumatol. 2010;37:1844–51.
26. Saito T, et al. Regional cerebral glucose metabolism in systemic lupus erythematosus patients with major depressive disorder. J Neurol Sci. 2017;379:127–30.
27. Katsumata Y, et al. Diagnostic reliability of cerebral spinal fluid tests for acute confusional state (delirium) in patients with systemic lupus erythematosus: interleukin 6 (IL-6), IL-8, interferon-alpha, IgG index, and Q-albumin. J Rheumatol. 2007;34:2010–7.
28. Panopalis P, et al. Impact of memory impairment on employment status in persons with systemic lupus erythematosus. Arthritis Rheum. 2007;57:1453–60.
29. Ad Hoc Committee on Lupus Response Criteria, Cognition Sub-committee, Mikdashi JA, et al. Proposed response criteria for neurocognitive impairment in systemic lupus erythematosus clinical trials. Lupus. 2007;16:418–25.
30. Kozora E, et al. Cognitive dysfunction in systemic lupus erythematosus: past, present, and future. Arthritis Rheum. 2008;58:3286–98.
31. Brown ES. Effects of glucocorticoids on mood, memory, and the hippocampus. Treatment and preventive therapy. Ann NY Acad Sci. 2009;1179:41–55.
32. Nishimura K, et al. Neurocognitive impairment in corticosteroid-naive patients with active systemic lupus erythematosus: a prospective study. J Rheumatol. 2015;42:441–8.

33. Filley CM. White matter: organization and functional relevance. Neuropsychol Rev. 2010;20:158–73.
34. Appenzeller S, et al. Neurometabolic changes in normal white matter may predict appearance of hyperintense lesions in systemic lupus erythematosus. Lupus. 2007;16:963–71.
35. Filley CM, et al. White matter microstructure and cognition in non-neuropsychiatric systemic lupus erythematosus. Cogn Behav Neurol. 2009;22:38–44.
36. Ramage AE, et al. Neuroimaging evidence of white matter inflammation in newly diagnosed systemic lupus erythematosus. Arthritis Rheum. 2011;63:3048–57.
37. Tomietto P, et al. General and specific factors associated with severity of cognitive impairment in systemic lupus erythematosus. Arthritis Rheum. 2007;57:1461–72.
38. Sanna G, et al. Neuropsychiatric manifestations in systemic lupus erythematosus: prevalence and association with antiphospholipid antibodies. J Rheumatol. 2003;30:985–92.
39. Grover S, et al. Catatonia in systemic lupus erythematosus: a case report and review of literature. Lupus. 2013;22:634–8.
40. Bica BE, et al. Electroconvulsive therapy as a treatment for refractory neuropsychiatric lupus with catatonia: three case studies and literature review. Lupus. 2015;24:1327–31.
41. Wang HY, Huang TL. Benzodiazepines in catatonia associated with systemic lupus erythematosus. Psychiatry Clin Neurosci. 2006;60:768–70.
42. Ishikura R, et al. Factors associated with anxiety, depression and suicide ideation in female outpatients with SLE in Japan. Clin Rheumatol. 2001;20:394–400.
43. Mok CC, et al. Suicidal ideation in patients with systemic lupus erythematosus: incidence and risk factors. Rheumatology. 2014;53:714–21.
44. Hajduk A, et al. Prevalence and correlates of suicidal thoughts in patients with neuropsychiatric lupus. Lupus. 2016;25:185–92.
45. Tang KT, et al. Suicidal drug overdose in patients with systemic lupus erythematosus, a nationwide population-based case-control study. Lupus. 2016;25:199–203.
46. Bhangle SD, et al. Corticosteroid-induced neuropsychiatric disorders: review and contrast with neuropsychiatric lupus. Rheumatol Int. 2013;33:1923–32.
47. Chau SY, Mok CC. Factors predictive of corticosteroid psychosis in patients with systemic lupus erythematosus. Neurology. 2003;61:104–7.
48. Nishimura K, et al. Blood-brain barrier damage as a risk factor for corticosteroid-induced psychiatric disorders in systemic lupus erythematosus. Psychoneuroendocrinology. 2008;33:395–403.
49. Nishimura K, et al. New-onset psychiatric disorders after corticosteroid therapy in systemic lupus erythematosus: an observational case-series study. J Neurol. 2014;261:2150–8.
50. Nishimura K, et al. Risperidone in the treatment of corticosteroid-induced mood disorders, manic/mixed episodes, in systemic lupus erythematosus: a case series. Psychosomatics. 2012;53:289–93.

Chapter 10
Treatment of Neuropsychiatric Systemic Lupus Erythematosus

Tetsuji Sawada

Abstract Systemic lupus erythematosus includes a wide range of neuropsychiatric manifestations, in which various autoantibodies and cytokines could play a pathophysiologic role. The key drugs used in the treatment of neuropsychiatric systemic lupus erythematosus include corticosteroids and immunosuppressive agents to suppress the underlying inflammatory pathology, antiplatelet and antithrombotic agents in the context of thrombosis mediated by antiphospholipid antibodies, and agents for symptomatic treatment of individual neuropsychiatric manifestations, such as antiepileptic drugs for seizures and antipsychotic drugs for lupus psychosis.

Keywords Treatment · Corticosteroids · Cyclophosphamide · Antiplatelet and anticoagulant drugs · Symptomatic treatment

10.1 Introduction

Systemic lupus erythematosus (SLE) is a chronic inflammatory disease that develops in genetically susceptible individuals in response to environmental factors, including viral infection and ultraviolet light. It is an autoimmune disease characterized by production of various autoantibodies and cytokines, such as antinuclear antibodies and interferon-alpha, leading to systemic inflammation and multiple organ damage and dysfunction. Central and peripheral nervous system involvement in SLE is known as neuropsychiatric SLE (NPSLE). This is one of the refractory manifestations of SLE, and its diagnosis and management are often challenging for clinicians [1].

A wide variety of neuropsychiatric manifestations occur in NPSLE. In terms of prevalence and severity, higher brain dysfunction and seizures are the central nervous system (CNS) symptoms most commonly encountered by health care provid-

T. Sawada (✉)
Department of Rheumatology, Tokyo Medical University Hospital, Tokyo, Japan
e-mail: tsawada@tokyo-med.ac.jp

ers involved in the management of patients with NPSLE. This chapter provides an overview of the clinical manifestations and treatment of NPSLE, including the use of immunosuppressive, antiplatelet, and antithrombotic drugs, as well as drugs for managing symptoms.

10.2 Pathogenesis of NPSLE

In 1999, the American College of Rheumatology published a classification of the neuropsychiatric manifestations of SLE [1], in which the CNS manifestations were classified into two domains: (i) neurologic syndromes characterized by regional brain symptoms and (ii) diffuse psychiatric neuropsychologic syndromes characterized by impairment of higher brain function (Table 10.1). The second domain, known as lupus psychosis, was further categorized into five subdomains: acute confusional state, anxiety disorder, cognitive dysfunction, mood disorder, and psychosis. Regarding the pathogenesis of the CNS manifestations of NPSLE, some of the neurologic syndromes, such as cerebrovascular disease, involve thrombosis caused

Table 10.1 Neuropsychiatric syndromes associated with systemic lupus erythematosus

Central nervous system
Psychiatric manifestations
Acute confusional state
Psychosis
Anxiety disorder
Mood disorder
Cognitive dysfunction
Neurologic syndromes
Aseptic meningitis
Cerebrovascular disease
Headache
Movement disorder (chorea)
Seizures
Demyelinating syndrome
Myelopathy (transverse myelitis)
Peripheral nervous system
Autonomic disorder
Mononeuropathy
Cranial neuropathy
Plexopathy
Polyneuropathy
Acute inflammatory demyelinating polyradiculoneuropathy (Guillain-Barré syndrome)
Myasthenia gravis

by antiphospholipid antibodies rather than inflammation affecting the nervous system via autoimmune mechanisms. Further, neurologic syndromes and diffuse psychiatric neuropsychologic syndromes may coexist. For example, seizures, which are categorized as neurologic syndromes, are often comorbid with lupus psychosis.

The severity of NPSLE is not necessarily reflected by the systemic activity of SLE. Thus, serum levels of anti-DNA antibodies and complement proteins, which are immune markers of the severity of lupus nephritis, are not correlated with the severity of NPSLE. Therefore, it can be difficult to evaluate patients with NPSLE by routine clinical examination. It has recently been demonstrated that autoantibodies, such as antiribosomal P protein antibodies, could play a key role in the development of lupus psychosis by binding to cell surface antigens and affecting the functioning of brain cells [2]. Further immunologic abnormalities in the CNS that have been implicated in the pathogenesis of lupus psychosis include production of certain cytokines, such as interleukin-6 and interferon-alpha [3–5]. Although the levels of some of these autoantibodies and cytokines are usually measured for research purposes, there are occasions when they should be measured as part of the evaluation of patients with lupus psychosis.

10.3 Treatment of NPSLE: Overviews

SLE has been demonstrated to be an independent risk factor for cardiovascular events, so it is important to control modifiable cardiovascular risk factors, including obesity, hypertension, dyslipidemia, smoking, diabetes mellitus, and physical inactivity, in patients with SLE regardless of the presence of NPSLE. Anti-platelet agents, such as low-dose aspirin or hydroxychloroquine, may be beneficial for primary prevention of cardiovascular disease in patients with SLE and antiphospholipid antibodies [6].

A multidisciplinary approach to the diagnosis and treatment of NPSLE in collaboration with relevant specialists, such as neurologists, psychiatrists, and neuroradiologists, is needed for patients with SLE who present with neuropsychiatric manifestations. First, the differential diagnosis, which includes infection and metabolic, electrolyte, iatrogenic, and hypertension-related disorders, should be considered. Once the diagnosis of NPSLE is established and classified according to the American College of Rheumatology nomenclature, it is necessary to identify the underlying pathology of the neuropsychiatric manifestations—that is, whether NPSLE is caused by inflammation, related to pro-inflammation and/or autoimmunity, or thrombosis/ischemia, often related to the presence of antiphospholipid antibodies with prothrombotic activity [7].

Three main classes of drugs are used to treat NPSLE: immunosuppressants that suppress inflammatory and autoimmune reactions occurring in the CNS; antiplatelet and anticoagulant medications that reduce the risk of thromboembolic events; and medications that treat symptoms, such as antiepileptic and antipsychotic drugs.

First, intense immunosuppressive treatment is indicated for patients with NPSLE in whom an autoimmune mechanism is identified as the cause of CNS damage. Measurement of interleukin-6 levels, the immunoglobulin G index, and antineuronal antibody levels in the cerebrospinal fluid (CSF) may provide evidence to support an autoimmunity-mediated mechanism [4, 8, 9], although it is still important to exclude other causes of neuropsychiatric manifestations. Intense immunosuppressive treatment with high-dose corticosteroids, often in combination with cyclophosphamide, is required in patients with SLE who present with acute neuropsychiatric manifestations, especially those with concomitant high systemic SLE activity. Neuropsychologic tests for higher brain function, electroencephalography, magnetic resonance imaging [10], computed tomography, single-photon emission computed tomography, and CSF analysis (including interleukin-6 levels) are useful for both diagnosis and monitoring in patients with NPSLE. Other immunosuppressive treatments, including plasma exchange, intravenous (IV) immunoglobulin, and rituximab, are used as alternative therapies in refractory cases. Second, some of the neurologic and psychologic manifestations of NPSLE, such as cerebrovascular disease, migraine-like headache, and cognitive dysfunction, have a thromboembolic etiology, and are therefore treated with antiplatelet and anticoagulant drugs, particularly in patients who are carriers of antiphospholipid antibodies. Third, symptomatic treatment is a crucial aspect of managing NPSLE. Therefore, antiepileptic and antipsychotic drugs may also be used [11].

10.4 Management of the Main Neuropsychiatric Manifestations

The mainstay of treatment for the CNS manifestations of NPSLE that are unrelated to thrombosis is immunosuppressive therapy, including corticosteroids and cyclophosphamide [12]. Cyclophosphamide was found to be superior to corticosteroids in the treatment of acute NPSLE and severe CNS and/or peripheral nervous system symptoms in a controlled clinical trial reported by Barile-Fabris et al. [13]. However, that trial only included 32 patients and no other clinical comparisons of the efficacy of these two agents have been reported since, so it is still unknown which of these agents is superior in the treatment of NPSLE. In routine clinical practice, a combination of high-dose corticosteroids and IV cyclophosphamide is used as induction therapy, followed by maintenance therapy using azathioprine or mycophenolate mofetil (MMF), as used in patients with lupus nephritis.

10.4.1 Acute Confusional State

Acute confusional state, previously known as acute organic brain syndrome or encephalopathy, is a severe manifestation of NPSLE, for which intense immunosuppression therapy is mandatory. A combination of high-dose corticosteroids and an immunosuppressant, such as cyclophosphamide, can be used to correct the underlying autoimmune abnormalities in the CNS, and is efficacious in many cases, with response rates of up to 70% [6]. Haloperidol or an atypical antipsychotic agent can be used as adjunctive symptomatic treatment for agitated delirium concomitantly with immunosuppressive agents. Plasma exchange therapy and rituximab are used in refractory cases.

10.4.2 Mood Disorder, Anxiety Disorder, and Psychosis

Patients with these disorders present with a range of neurologic and psychiatric symptoms that vary in severity. Patients with severe symptoms are treated with intense immunosuppression therapy similar to that used in patients with acute confusional state. However, psychiatric medications alone may suffice in patients with mild psychiatric symptoms, especially in the setting of low systemic disease activity without severe damage to internal organs [11].

10.4.3 Cognitive Dysfunction

Mild to moderate cognitive dysfunction is common in patients with SLE, with a reported frequency of 14–79% [14], whereas severe cognitive impairment is rare (3–5%) [6]. The causes of cognitive dysfunction in NPSLE include hypertension, brain lesions (atrophy and multiple cerebral infarctions seen on magnetic resonance images) accumulation of damage following onset of lupus, and the presence of antiphospholipid antibodies [15]. Therefore, cardiovascular risk factors should be controlled in patients with SLE, starting as soon as the diagnosis is made. An association of persistent elevation of anticardiolipin antibodies with poor cognitive function has been demonstrated in patients with SLE and antiphospholipid antibodies [16, 17]. Further, McLaurin et al. reported that regular use of aspirin was associated with improved cognitive function in older patients with SLE and other risk factors for vascular disease [18]. However, no relevant controlled clinical trials have been performed, so the role of antiplatelet and antithrombotic drugs in the prevention of cognitive functional decline in patients with SLE has yet to be firmly established.

A study of the relationship between cognitive dysfunction and CNS inflammation in patients with SLE by Denburg et al. found that brief exposure to relatively low doses of corticosteroids (prednisone 0.5 mg/kg daily) improved cognition and mood in 5 of 8 women with mild SLE in an n-of-1 double-blind controlled trial [19]. However, no further randomized controlled trials demonstrating the effects of corticosteroids on CNS function in SLE have been reported. Therefore, use of corticosteroids to treat cognitive dysfunction should be considered only for patients with high disease activity. Memantine, an N-methyl D-aspartate (NMDA) receptor antagonist used as a symptomatic and neuroprotective treatment in patients with Alzheimer's disease, was reported to be ineffective when used to improve cognitive function in SLE [20].

10.4.4 Seizures

The prevalence of seizure disorders in patients with SLE ranges from 7% to 40%, with an average of 15% [21], which seems to be higher than that in the general population. Seizure disorders, which are classified as neurologic CNS syndromes, often coexist with diffuse psychiatric neuropsychologic syndromes in patients with NPSLE. Antineuronal antibodies, anti-NMDA receptor subunit NR2 antibodies, and antiribosomal P protein antibodies have been detected in the CSF of patients with SLE and seizures complicated by psychosis [22–24]. Therefore, patients with NPSLE and high systemic disease activity who develop seizures are treated with a combination of immunosuppressive therapy and symptomatic antiepileptic drugs. Immunosuppressive treatment alone can control seizures without the need for antiepileptic drugs in some cases. Of note, seizure disorders may be related to comorbidities such as uremia, hypertension, CNS infection, and stroke, which should be controlled adequately. Antiphospholipid antibodies might be found in seizure disorders in the absence of cerebrovascular diseases. Cessation of antiepileptic drugs is possible in some patients when SLE activity and comorbid conditions are controlled. Recurrence of mild seizures is often controlled by antiepileptic drugs alone.

10.4.5 Headache

Headache is a common manifestation of SLE and is classified as one of the neurologic syndromes affecting the CNS in these patients. However, it has been suggested that the prevalence, phenotype, and treatment of headache in patients with SLE are not significantly different from those in the general population [25]. Moreover, headaches are not associated with disease activity in SLE, except possibly for lupus headache, defined as a severe, persistent headache that may be migrainous, but must be nonresponsive to narcotic analgesia [26]; however, it is rare (1.5%) and the term is not universally accepted [27]. In general, headache in patients with SLE is treated in much the same way as primary headache [25].

10.4.6 Movement Disorder

Movement disorder is a rare neurologic manifestation of NPSLE, and is thought to involve antiphospholipid and antineuronal antibodies [28, 29], indicating that control of the underlying autoimmune mechanisms could help to ameliorate involuntary movements. Treatments for movement disorder associated with SLE include dopamine antagonists as symptomatic therapy, corticosteroids with or without immunosuppressive drugs, and antithrombotic drugs for patients with antiphospholipid antibodies.

10.4.7 Cerebrovascular Disease

Cerebrovascular events, such as ischemic stroke, are severe CNS manifestations of NPSLE. The management of acute cerebrovascular disease in patients with SLE is similar to that in the general population. When cerebrovascular disease occurs in the setting of high disease activity, corticosteroids with/without immunosuppressive agents are used to suppress overall SLE activity. When acute cerebrovascular disease develops in the context of persistently positive antiphospholipid antibodies, secondary prevention measures are necessary [30]; however, the optimal management remains controversial, especially with respect to the relative merits of antiplatelet agents and warfarin. In general, the first-line strategy for prevention of non-cardioembolic stroke in high-risk patients (triple positive antiphospholipid antibody profile, multiple lesions seen on magnetic resonance imaging of the brain, history of arterial clot, active SLE, and smoking) is a combination of antiplatelet agents (aspirin plus clopidogrel or ticagrelor), warfarin with a target international ratio (INR) of 2.0–3.0 plus aspirin, or high-dose warfarin with a target INR of 3.0–4.0. Aspirin or a combination of antiplatelet agents is used in low-risk patients, although some experts also recommend warfarin with a target INR of 2.0–3.0 [31].

10.5 Immune-Modulating Drugs Used for NPSLE

10.5.1 Corticosteroids

Since the 1950s, corticosteroids have been considered the most effective agents available for the treatment of a number of severe clinical manifestations of SLE, including neuropsychiatric symptoms [32–34]. In general, corticosteroid doses are categorized as low, medium, high, very high, or pulsed, and the choice between them is determined by the severity of organ damage in patients with SLE [35]. Given that NPSLE is a major cause of morbidity and mortality in SLE [36–38], corticosteroids are administered in high doses as first-line treatment for NPSLE, either by oral administration at doses equivalent to prednisolone 1 mg/kg/day or by pulse IV administration of methylprednisolone 1000 mg, usually for three successive days followed by switching to the oral route. High-dose corticosteroid therapy is continued for 2–4 weeks until disease activity is controlled, and the dose is tapered gradually thereafter according to the clinical and immunologic SLE activity.

Although corticosteroids remain the mainstay of treatment for SLE [39], they have many side effects, including psychosis, infection, peptic ulcer, osteoporosis, hypertension, hyperglycemia, hypercholesterolemia, avascular necrosis, myopathy, cataract, and glaucoma. Steroid-induced psychosis is a dose-dependent adverse reaction, and a prospective study of 32 patients with asthma who received 40 mg or more of prednisolone for at least a week demonstrated a significant increase in manic symptoms during the first 3–7 days of prednisone therapy [40]. It should be noted that lupus psychosis can emerge de novo or as a result of deterioration in patients with a history of NPSLE after initiation of high-dose corticosteroid therapy. Of interest, Shimizu et al. recently conducted a retrospective study involving 146 patients with SLE and 162 patients with other systemic autoimmune diseases who were consecutively recruited and treated with prednisolone at a dose of 40 mg/day or more, and demonstrated that the prevalence of post-steroid neuropsychiatric symptoms was significantly higher in patients with SLE than in the controls (24.7% versus 7.4%). This finding suggests that the population with NPSLE is likely to include patients with post-steroid neuropsychiatric symptoms [41]. It has been suggested that lupus psychosis and steroid-induced psychosis are not mutually exclusive and that both conditions can occur in the same individual [42].

10.5.2 Cyclophosphamide

Cyclophosphamide is an alkylating agent with immunosuppressive effects and is used to treat autoimmune diseases and systemic vasculitis. Although cyclophosphamide can be administered via several routes, it is generally administered IV (according to the standard protocol of the US National Institutes of Health or its variations), mainly because of the difference in exposure between the oral and IV routes [43]. Cyclophosphamide in combination with corticosteroids is the most commonly used immunosuppressive therapy in patients with severe NPSLE [44]. The efficacy of cyclophosphamide in NPSLE has been documented in many case reports and case series [45–53].

Barile-Fabris et al. compared the efficacy of IV cyclophosphamide and IV methylprednisolone in a controlled clinical trial that included 32 patients with NPSLE who had severe neuropsychiatric manifestations, such as seizures, optic neuritis, peripheral or cranial neuropathy, coma, brainstem disease, and transverse myelitis [13]. In this study, all the patients initially received methylprednisolone 1 g/day for 3 days as induction therapy followed by oral prednisone 1 mg/kg/day starting on day 4 for up to 3 months and tapered thereafter according to disease status. The patients randomized to methylprednisolone received 1 g/day for 3 days every month for 4 months, then twice monthly for 6 months, and then every 3 months for a year, while those randomized to cyclophosphamide received 0.75 g/m^2 body surface area monthly for a year and then every 3 months for a further year. The response rate was significantly higher in the cyclophosphamide group than in the methylprednisolone therapy (95% vs. 54%, p = 0.03). Although the results of that study suggested that cyclophosphamide is superior to high-dose corticosteroids for the treatment of severe NPSLE, the number of patients included was small, and further well-designed studies with defined outcome measures in larger numbers of patients are needed to confirm the advantage of cyclophosphamide over methylprednisolone in NPSLE.

10.5.3 Hydroxychloroquine

Antimalarials, such as hydroxychloroquine, have long been used as immunomodulatory agents for SLE. Multi-regression analysis in the multiethnic lupus cohort (LUMINA, Lupus in minorities: Nature versus Nurture) study that included 632 patients with SLE, of whom 185 developed NPSLE, revealed that hydroxychloroquine and a moderate dose of prednisolone delayed the onset of the first manifestation of NPSLE, indicating a protective effect of hydroxychloroquine [54].

10.5.4 Intravenous Immunoglobulin

Intravenous (IV) immunoglobulin has been used to treat a wide range of clinical manifestations of SLE, although the level of evidence is anecdotal [55, 56], and has been used experimentally as induction therapy for refractory acute NPSLE [57–59].

10.5.5 Plasmapheresis

Plasmapheresis and immunoadsorption have been used in the management of SLE, based on the assumption that removal of autoantibodies, activated complement components, coagulation factors, cytokines, and microparticles would be beneficial in patients with the disease. Observational case series demonstrating the beneficial effects of plasmapheresis for NPSLE either alone or synchronized with IV cyclophosphamide have also been reported [60, 61].

10.5.6 Mycophenolate Mofetil (MMF)

MMF is an immunosuppressive agent that blocks the de novo pathway of purine synthesis by inhibiting inosine monophosphate dehydrogenase. MMF is an established first-line immunosuppressant used for induction of remission as well as maintenance therapy in lupus nephritis [62, 63]. Tselios et al. investigated the effectiveness of MMF in patients with and without renal involvement, and demonstrated that MMF could also be an efficacious alternative to standard medications in patients with non-renal manifestations of SLE [64]. MMF in combination with corticosteroids was shown to induce complete remission of the CNS manifestations of SLE, including lupus headache and acute confusional state, in 14 of 18 patients (78%) after 12 months. Further, MMF was found to have modest to moderate clinical efficacy in 3 patients with NPSLE in the post hoc analysis of ALMS (the Aspreva Lupus Management Study) [65], with similar findings in a retrospective cohort study (n = 6) by Conti et al. [66] and in case reports [67–69], and case series [70, 71]. However, there is as yet no evidence from randomized controlled trials which has demonstrated the efficacy of the use of MMF in NPSLE. Therefore, MMF is considered to have a limited role in the treatment of NPSLE [72], and is not included as a treatment option in the European League Against Rheumatism recommendations for management of NPSLE [6].

10.5.7 Azathioprine

Azathioprine is a purine analog that acts as an antimetabolite immunosuppressive drug and has traditionally been used as maintenance therapy for lupus nephritis [73] and for the treatment of nonrenal manifestations of SLE, although the evidence supporting its use is limited. The literature on the use of azathioprine as maintenance treatment for severe NPSLE consists only of case series [74, 75].

10.5.8 Rituximab

Rituximab is a chimeric anti-CD20 monoclonal antibody with human IgG1 constant domains. Observational case studies have demonstrated the clinical efficacy of rituximab in the treatment of NPSLE [76–78]. Tokunaga et al. reported that rituximab led to rapid improvement of CNS-related manifestations, particularly acute confusional state, in 10 of 10 patients with NPSLE refractory to intensive standard treatment [76]. Furthermore, in the systematic review of case reports and open-label studies involving 35 patients with refractory NPSLE, Narváez et al. found that a complete or partial therapeutic response was achieved in 85% of patients after one cycle of treatment, although relapse occurred in 45% of patients after cessation of rituximab and infections occurred in 29% of cases [77].

10.5.9 Belimumab

Belimumab is a human IgG1 monoclonal antibody that reacts with B-lymphocyte stimulator, also known as BAFF (B cell-activating factor belonging to the TNF family). In the BLISS (Belimumab in Subjects With Systemic Lupus

Erythematosus)-52 and BLISS-76 studies, patients with severe CNS manifestations were excluded [79, 80]. However, post hoc analysis of 45 patients with SLE and CNS manifestations, including 24 patients with lupus headache, in BLISS-52 and BLISS-76 revealed response rates in the groups allocated to placebo, belimumab 1 mg/kg, and belimumab 10 mg/kg of 20.0%, 100%, and 69.2%, respectively, indicating that belimumab ameliorates lupus headache, albeit modestly [81].

10.6 Summary

NPSLE is the most frequent and most intractable manifestation of CNS abnormalities in patients with autoimmune connective tissue disease [82]. There were times in the past when the etiology of NPSLE was ambiguously considered to be vasculitis of the CNS, and no effective diagnostic tools or treatments were available. However, it is now known that various autoantibodies and cytokines play a key role in the pathogenesis of CNS abnormalities in NPSLE. Further investigations of the mechanism of action leading to these immunologic aberrations are needed to enable development of new treatments targeting them.

References

1. ACR Ad Hoc Committee on Neuropsychiatric Lupus Nomenclature. The American College of Rheumatology nomenclature and case definitions for neuropsychiatric lupus syndromes. Arthritis Rheum. 1999;42:599–608.
2. Matus S, et al. Antiribosomal-P autoantibodies from psychiatric lupus target a novel neuronal surface protein causing calcium influx and apoptosis. J Exp Med. 2007;204:3221–34.
3. Shiozawa S, et al. Interferon-alpha in lupus psychosis. Arthritis Rheum. 1992;35:417–22.
4. Hirohata S, et al. Accuracy of cerebrospinal fluid IL-6 testing for diagnosis of lupus psychosis. A multicenter retrospective study. Clin Rheumatol. 2009;28:1319–23.
5. Yoshio T, et al. IL-6, IL-8, IP-10, MCP-1 and G-CSF are significantly increased in cerebrospinal fluid but not in sera of patients with central neuropsychiatric lupus erythematosus. Lupus. 2016;25:997–1003.
6. Bertsias GK, et al. EULAR recommendations for the management of systemic lupus erythematosus with neuropsychiatric manifestations: report of a task force of the EULAR standing committee for clinical affairs. Ann Rheum Dis. 2010;69:2074–82.
7. Magro-Checa C, et al. Management of neuropsychiatric systemic lupus erythematosus: current approaches and future perspectives. Drugs. 2016;76:459–83.
8. Hirohata S, Hirose S, Miyamoto T. Cerebrospinal fluid IgM, IgA, and IgG indexes in systemic lupus erythematosus. Their use as estimates of central nervous system disease activity. Arch Intern Med. 1985;145:1843–6.
9. Iizuka N, et al. Identification of autoantigens specific for systemic lupus erythematosus with central nervous system involvement. Lupus. 2010;19:717–26.
10. Arinuma Y, et al. Brain MRI in patients with diffuse psychiatric/neuropsychological syndromes in systemic lupus erythematosus. Lupus Sci Med. 2014;1:e000050.
11. Nishimura K, et al. Risperidone in the treatment of acute confusional state (delirium) due to neuropsychiatric lupus erythematosus: case report. Int J Psychiatry Med. 2003;33:299–303.

12. Fernandes Moca Trevisani V, Castro AA, Ferreira Neves Neto J, Atallah AN. Cyclophosphamide versus methylprednisolone for treating neuropsychiatric involvement in systemic lupus erythematosus. Cochrane Database Syst Rev. 2013; CD002265.
13. Barile-Fabris L, et al. Controlled clinical trial of IV cyclophosphamide versus IV methylprednisolone in severe neurological manifestations in systemic lupus erythematosus. Ann Rheum Dis. 2005;64:620–5.
14. Kozora E, Filley CM. Cognitive dysfunction and white matter abnormalities in systemic lupus erythematosus. J Int Neuropsychol Soc. 2011;17:385–92.
15. Tomietto P, et al. General and specific factors associated with severity of cognitive impairment in systemic lupus erythematosus. Arthritis Rheum. 2007;57:1461–72.
16. Menon S, et al. A longitudinal study of anticardiolipin antibody levels and cognitive functioning in systemic lupus erythematosus. Arthritis Rheum. 1999;42:735–41.
17. Hanly JG, et al. A prospective analysis of cognitive function and anticardiolipin antibodies in systemic lupus erythematosus. Arthritis Rheum. 1999;42:728–34.
18. McLaurin EY, et al. Predictors of cognitive dysfunction in patients with systemic lupus erythematosus. Neurology. 2005;64:297–303.
19. Denburg SD, et al. Corticosteroids and neuropsychological functioning in patients with systemic lupus erythematosus. Arthritis Rheum. 1994;37:1311–20.
20. Petri M, et al. Memantine in systemic lupus erythematosus: a randomized, double-blind placebo-controlled trial. Semin Arthritis Rheum. 2011;41:194–202.
21. Devinsky O, et al. Epilepsy associated with systemic autoimmune disorders. Epilepsy Cur. 2013;13:62–8.
22. Bluestein HG, et al. Cerebrospinal fluid antibodies to neuronal cells: association with neuropsychiatric manifestations of systemic lupus erythematosus. Am J Med. 1981;70:240–6.
23. Arinuma Y, et al. Association of cerebrospinal fluid anti-NR2 glutamate receptor antibodies with diffuse neuropsychiatric systemic lupus erythematosus. Arthritis Rheum. 2008;58:1130–5.
24. Yoshio T, et al. Antiribosomal P protein antibodies in cerebrospinal fluid are associated with neuropsychiatric systemic lupus erythematosus. J Rheumatol. 2005;32:34–9.
25. Mitsikostas DD, et al. A meta-analysis for headache in systemic lupus erythematosus: the evidence and the myth. Brain. 2004;127(Pt 5):1200–9.
26. Gladman DD, et al. Systemic lupus erythematosus disease activity index 2000. J Rheumatol. 2002;29:288–91.
27. Hanly JG, et al. Headache in systemic lupus erythematosus: results from a prospective, international inception cohort study. Arthritis Rheum. 2013;65:2887–97.
28. Dale RC, et al. Antibody binding to neuronal surface in movement disorders associated with lupus and antiphospholipid antibodies. Dev Med Child Neurol. 2011;53:522–8.
29. Krakauer M, Law I. FDG PET brain imaging in neuropsychiatric systemic lupus erythematosis with choreic symptoms. Clin Nucl Med. 2009;34:122–3.
30. Bala MM, et al. Antiplatelet and anticoagulant agents for secondary prevention of stroke and other thromboembolic events in people with antiphospholipid syndrome. Cochrane Database Syst Rev. 2017;10:CD012169.
31. Fonseca AG, D'Cruz DP. Controversies in the antiphospholipid syndrome: can we ever stop warfarin? J Autoimmune Dis. 2008;5:6.
32. Carey RA, et al. The effect of adrenocorticotropic hormone (ACTH) and cortisone on the course of disseminated lupus erythematosus and peri-arteritis nodosa. Bull Johns Hopkins Hosp. 1950;87:425–60.
33. Pickering G, et al. Treatment of systemic lupus erythematosus with steroids: report to the Medical Research Council by the collagen diseases and hypersensitivity panel. Br Med J. 1961;2:915–20.
34. Albert DA, et al. Does corticosteroid therapy affect the survival of patients with systemic lupus erythematosus? Arthritis Rheum. 1979;22:945–53.
35. Buttgereit F, et al. Standardised nomenclature for glucocorticoid dosages and glucocorticoid treatment regimens: current questions and tentative answers in rheumatology. Ann Rheum Dis. 2002;61:718–22.

36. Zhen J, et al. Death-related factors of systemic lupus erythematosus patients associated with the course of disease in Chinese populations: multicenter and retrospective study of 1,958 inpatients. Rheumatol Int. 2013;33(6):1541.
37. Jonsen A, et al. Outcome of neuropsychiatric systemic lupus erythematosus within a defined Swedish population: increased morbidity but low mortality. Rheumatology. 2002;41:1308–12.
38. Rubin LA, et al. Mortality in systemic lupus erythematosus: the bimodal pattern revisited. Q J Med. 1985;55:87–98.
39. Kasturi S, Sammaritano LR. Corticosteroids in Lupus. Rheum Dis Clin N Am. 2016;42:47–62,viii
40. Brown ES, et al. Mood changes during prednisone bursts in outpatients with asthma. J Clin Psychopharmacol. 2002;22:55–61.
41. Shimizu Y, et al. Post-steroid neuropsychiatric manifestations are significantly more frequent in SLE compared with other systemic autoimmune diseases and predict better prognosis compared with de novo neuropsychiatric SLE. Autoimmun Rev. 2016;15:786–94.
42. Hirohata S, et al. A patient with systemic lupus erythematosus presenting both central nervous system lupus and steroid induced psychosis. J Rheumatol. 1988;15:706–10.
43. Hebert LA, Rovin BH. Oral cyclophosphamide is on the verge of extinction as therapy for severe autoimmune diseases (especially lupus): should nephrologists care? Nephron Clin Pract. 2011;117:c8–14.
44. Kivity S, et al. Neuropsychiatric lupus: a mosaic of clinical presentations. BMC Med. 2015;13:43.
45. Boumpas DT, et al. Pulse cyclophosphamide for severe neuropsychiatric lupus. Q J Med. 1991;81:975–84.
46. Neuwelt CM, et al. Role of intravenous cyclophosphamide in the treatment of severe neuropsychiatric systemic lupus erythematosus. Am J Med. 1995;98:32–41.
47. Barile L, Lavalle C. Transverse myelitis in systemic lupus erythematosus–the effect of IV pulse methylprednisolone and cyclophosphamide. J Rheumatol. 1992;19:370–2.
48. Baca V, et al. Favorable response to intravenous methylprednisolone and cyclophosphamide in children with severe neuropsychiatric lupus. J Rheumatol. 1999;26:432–9.
49. Martin-Suarez I, et al. Immunosuppressive treatment in severe connective tissue diseases: effects of low dose intravenous cyclophosphamide. Ann Rheum Dis. 1997;56:481–7.
50. Ramos PC, et al. Pulse cyclophosphamide in the treatment of neuropsychiatric systemic lupus erythematosus. Clin Exp Rheumatol. 1996;14:295–9.
51. Stojanovich L, et al. Neuropsychiatric lupus favourable response to low dose i.v. cyclophosphamide and prednisolone (pilot study). Lupus. 2003;12:3–7.
52. Hirano T, et al. Successful therapy with steroid and cyclophosphamide pulse for CNS lupus and lupus myelitis. Jap J Clin Immunol. 2007;30:414–8.
53. Fanouriakis A, et al. Cyclophosphamide in combination with glucocorticoids for severe neuropsychiatric systemic lupus erythematosus: a retrospective, observational two-centre study. Lupus. 2016;25:627–36.
54. Gonzalez LA, et al. Time to neuropsychiatric damage occurrence in LUMINA (LXVI): a multi-ethnic lupus cohort. Lupus. 2009;18:822–30.
55. Levy Y, et al. A study of 20 SLE patients with intravenous immunoglobulin--clinical and serologic response. Lupus. 1999;8:705–12.
56. Sakthiswary R, D'Cruz D. Intravenous immunoglobulin in the therapeutic armamentarium of systemic lupus erythematosus: a systematic review and meta-analysis. Medicine. 2014;93:e86.
57. Toubi E, et al. High-dose intravenous immunoglobulins: an option in the treatment of systemic lupus erythematosus. Hum Immunol. 2005;66:395–402.
58. Milstone AM, et al. Treatment of acute neuropsychiatric lupus with intravenous immunoglobulin (IVIG): a case report and review of the literature. Clin Rheumatol. 2005;24:394–7.
59. Engel G, van Vollenhoven RF. Treatment of severe CNS lupus with intravenous immunoglobulin. J Clin Rheumatol. 1999; 5:228–32. Sherer Y, et al. Successful treatment of systemic lupus erythematosus cerebritis with intravenous immunoglobulin. Clin Rheumatol. 1999; 18:170–3.

60. Neuwelt CM. The role of plasmapheresis in the treatment of severe central nervous system neuropsychiatric systemic lupus erythematosus. Ther Apher Dial. 2003;7:173–82.
61. Bartolucci P, et al. Adjunctive plasma exchanges to treat neuropsychiatric lupus: a retrospective study on 10 patients. Lupus. 2007;16:817–22.
62. Appel GB, et al. Mycophenolate mofetil versus cyclophosphamide for induction treatment of lupus nephritis. J Am Soc Nephrol. 2009;20:1103–12.
63. Dooley MA, et al. Mycophenolate versus azathioprine as maintenance therapy for lupus nephritis. N Engl J Med. 2011;365:1886–95.
64. Tselios K, et al. Mycophenolate Mofetil in nonrenal manifestations of systemic lupus erythematosus: an observational cohort study. J Rheumatol. 2016;43:552–8.
65. Ginzler EM, et al. Nonrenal disease activity following mycophenolate mofetil or intravenous cyclophosphamide as induction treatment for lupus nephritis: findings in a multicenter, prospective, randomized, open-label, parallel-group clinical trial. Arthritis Rheum. 2010;62:211–21.
66. Conti F, et al. Mycophenolate mofetil in systemic lupus erythematosus: results from a retrospective study in a large monocentric cohort and review of the literature. Immunol Res. 2014;60:270–6.
67. Lhotta K, et al. Cerebral vasculitis in a patient with hereditary complete C4 deficiency and systemic lupus erythematosus. Lupus. 2004;13:139–41.
68. Jose J, et al. Mycophenolate mofetil in neuropsychiatric systemic lupus erythematosus. Indian J Med Sci. 2005;59:353–6.
69. Higashioka K, et al. Successful treatment of lupus cerebrovascular disease with mycophenolate mofetil. Intern Med. 2015;54:2255–9.
70. Mok CC, et al. Mycophenolate mofetil for lupus related myelopathy. Ann Rheum Dis. 2006;65:971–3.
71. Saison J, et al. Systemic lupus erythematosus-associated acute transverse myelitis: manifestations, treatments, outcomes, and prognostic factors in 20 patients. Lupus. 2015;24:74–81.
72. Gordon C, Amissah-Arthur MB, Gayed M, Brown S, Bruce IN, D'Cruz D, et al. The British Society for Rheumatology guideline for the management of systemic lupus erythematosus in adults. Rheumatology. 2018;57:e1–e45.
73. Feng L, et al. Mycophenolate mofetil versus azathioprine as maintenance therapy for lupus nephritis: a meta-analysis. Nephrology. 2013;18:104–10.
74. Mok CC, et al. Treatment of lupus psychosis with oral cyclophosphamide followed by azathioprine maintenance: an open-label study. Am J Med. 2003;115:59–62.
75. Ginzler E, et al. Long-term maintenance therapy with azathioprine in systemic lupus erythematosus. Arthritis Rheum. 1975;18:27–34.
76. Tokunaga M, et al. Efficacy of rituximab (anti-CD20) for refractory systemic lupus erythematosus involving the central nervous system. Ann Rheum Dis. 2007;66:470–5.
77. Narvaez J, Rios-Rodriguez V, de la Fuente D, Estrada P, Lopez-Vives L, Gomez-Vaquero C, et al. Rituximab therapy in refractory neuropsychiatric lupus: current clinical evidence. Semin Arthritis Rheum. 2011;41:364–72.
78. Dale RC, et al. Utility and safety of rituximab in pediatric autoimmune and inflammatory CNS disease. Neurology. 2014;83:142–50.
79. Furie R, et al. A phase III, randomized, placebo-controlled study of belimumab, a monoclonal antibody that inhibits B lymphocyte stimulator, in patients with systemic lupus erythematosus. Arthritis Rheum. 2011;63:3918–30.
80. Navarra SV, et al. Efficacy and safety of belimumab in patients with active systemic lupus erythematosus: a randomised, placebo-controlled, phase 3 trial. Lancet. 2011;377:721–31.
81. Manzi S, et al. Effects of belimumab, a B lymphocyte stimulator-specific inhibitor, on disease activity across multiple organ domains in patients with systemic lupus erythematosus: combined results from two phase III trials. Ann Rheum Dis. 2012;71:1833–8.
82. Hirohata S. 110th scientific meeting of the Japanese society of internal medicine: invited lecture: 5. Neurological involvement in connective tissue diseases. Nippon Naika Gakkai Zasshi. 2013;102:2214–23.

Chapter 11
Promising Treatment Alternatives

Taku Yoshio and Hiroshi Okamoto

Abstract The current therapeutic approach to the difficult manifestations of neuropsychiatric systemic lupus erythematosus (NPSLE) remains empirical and is based on clinical experience. Available data on the use of rituximab in refractory NPSLE come from a large number of case reports and some open-label studies. Two patients with persistently active NPSLE, despite conventional therapy, responded dramatically to rituximab are described in this chapter. Current evidence on the therapeutic use of rituximab in this chapter is also analyzed through the English-language literatures. Evidence for the effectiveness of rituximab as induction therapy in NPSLE is based solely on several case reports and non-controlled trials. Although it is not yet possible to make definite recommendations, the global analysis of these cases supports the off-label use of rituximab in cases of severe refractory NPSLE. Furthermore, we present the blockade of new targets which may impact the future treatment of NPSLE.

Keywords NPSLE · Anti-CD20 antibodies · Rituximab · Anti-CD22 antibodies · Epratuzumab · Anti-BLyS antibodies · Belimumab · Tabalumab · Blisibimod · BLyS and april receptor · Atacicept · Anti-IL-6 receptor antibodies · Tocilizumab · Anti-IFN-α antibodies · Sifalimumab · Rontalizumab · Anti-terminal complement components C5a and C5b-9 antibodies · Eculizumab

---

T. Yoshio (✉)
Division of Rheumatology and Clinical Immunology, School of Medicine, Jichi Medical University, Shimotsuke-shi, Tochigi, Japan
e-mail: takuyosh@jichi.ac.jp

H. Okamoto
Minami-otsuka institute of technology, Minami-otsuka Clinic, Tokyo, Japan

© Springer International Publishing AG, part of Springer Nature 2018
S. Hirohata (ed.), *Neuropsychiatric Systemic Lupus Erythematosus*,
https://doi.org/10.1007/978-3-319-76496-2_11

11.1 Introduction

The involvement of the central nervous system (CNS) is one of the major causes of morbidity and mortality in patients with systemic lupus erythematosus (SLE), and it is the least understood aspect of the disease [1]. Indeed, its recognition and treatment continue to represent a major diagnostic and therapeutic challenge. Due to the lack of controlled randomized trials, the current therapeutic approach to the different manifestations of neuropsychiatric SLE (NPSLE) remains empirical and is based on clinical experience [2, 3]. Various combinations of corticosteroids, immunosuppressants (cyclophosphamide, mycophenolate mofetil, methotrexate and cyclosporine A), anticoagulant therapy, intravenous immunoglobulins and therapeutic plasma exchange are used depending on the presumptive main pathogenic mechanism, although resistant cases have been described.

On the other hand, the emergence of B-cell-depleting therapy with the monoclonal antibody rituximab, directed against the B-cell-specific antigen CD20, could increase the therapeutic armamentarium with respect to SLE [4]. In recent years a considerable number of observational studies and case reports have demonstrated encouraging early results with rituximab in cases of severe refractory NPSLE [5, 6], although its use is limited by the lack of licensing. In this chapter, we present 2 patients with persistently active NPSLE (case 1: mutism and acute confusional state, case 2: grand mal seizures and psychosis), despite conventional therapy, who responded dramatically to rituximab. We also review current evidence on the therapeutic use of rituximab in adult patients. Furthermore, we present the blockade of new targets which may impact future treatment of NPSLE.

11.2 Anti-CD20 Antibodies (Rituximab)

Rituximab is a chimeric monoclonal antibody directed against the B-cell-specific antigen CD20. B-cell depletion is achieved by rituximab through different mechanisms such as antibody-dependent cell-mediated cytotoxicity, complement-dependent cytotoxicity, and induction of apoptosis (Fig. 11.1).

11.2.1 Clinical Trials and Cohort Studies of Rituximab

The use of B-cell depletion therapy in SLE is based on the aspect that B cells play a central role in the pathogenesis of SLE, as antigen-presenting cells and in the production of autoantibodies, cytokines, and chemokines. Currently, rituximab is widely used as an alternative therapy in patients with active SLE who are nonresponsive to standard immunosuppressive therapy [7].

Two randomized controlled trials (RCTs) on rituximab in SLE patients, one with renal involvement [the Lupus Nephritis Assessment with Rituximab (LUNAR)

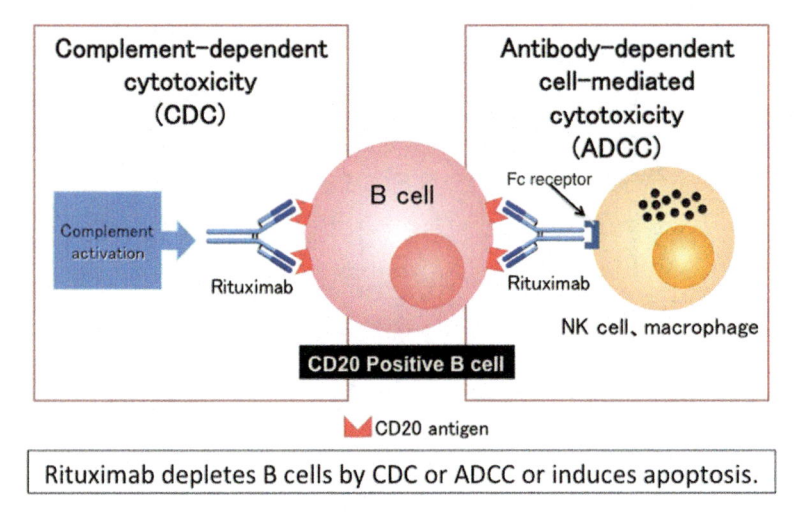

Fig. 11.1 The action of rituximab against B cells

trial] and the other without renal involvement [the Exploratory Phase II/III SLE Evaluation of Rituximab (EXPLORER) trial], failed to find superiority of rituximab over standard immunosuppressive regimens (glucocorticoids, cyclophosphamide, and mycophenolate mofetil) in mild to moderately active SLE. In these studies, however, severe and/or refractory patients were not included [8, 9]. Therefore, it might be difficult to find superiority of rituximab over the placebo in the presence of standard immunosuppressive regimens (glucocorticoids, cyclophosphamide and mycophenolate mofetil).

On the other hand, the efficacy and safety of rituximab in the treatment of nonrenal SLE has recently been analyzed in a systematic review including one RCT, two open-label studies and 22 cohort studies, with a total of 1231 patients. Rituximab was shown to be safe and effective in the treatment of nonrenal SLE, especially in terms of disease activity, immunologic parameters, and corticosteroid-sparing effect [10]. However, the efficacy of rituximab in NPSLE was not assessed in these studies mentioned above.

11.2.2 Regimen and Efficacy of Rituximab in Patients with NPSLE

In a substantial number of case reports and some open-label studies, good efficacy of rituximab was shown in refractory NPSLE [11–26]. The numbers of patients treated with rituximab, dose of rituximab, outcome in NPSLE patients excluding Japanese patients and in Japanese patients with NPSLE are shown in Table 11.1 and Table 11.2, respectively.

Table 11.1 The summary of NPSLE patients treated with rituximab (excluding Japanese patients)

	Authors [Ref No.]	Publish year	Article	Patient No.	RTX dose per once	Treated No.	Duration (weeks)	Outcome Complete response	Partial response	Non responder
1	Weide et al. [11]	2003	Case report	1	375 mg/m^2	4	1	1	0	0
2	Gottenberg et al. [12]	2004	Retrospective study	1	375 mg/m^2	4	1	1	0	0
3	Leandro et al. [13]	2005	I/II open study	3	1000 mg/m^2	2	2	1	1	1 (Outcome unknown)
4	Armstrong et al. [14]	2006	Case report	1	500 mg/m^2	2	2	1	0	0
5	Birnbaum et al. [15]	2008	Case report	1	1000 mg/m^2	2	2	1 (Remission)	0	0
6	Abud-Mendoza et al. [16]	2009	Prospective study	6	500 or 1000 mg/m^2	1 or 2	2	6	0	0
7	Lu et al. [17]	2009	Retrospective study	2	500 or 1000 mg/m^2	2	2	2	0	0
8	Nasir et al. [18]	2009	Case report	1	1000 mg/m^2	2	2	0	0	1
9	Espinosa et al. [19]	2010	Case report	2	375 mg/m^2	4	1	2	0	0
10	Nárvaez et al. [20]	2011	Case report	1	1000 mg/m^2	2	2	1	0	0
11	Pinto et al. [21]	2011	Retrospective open study	6	1000 mg/m^2	2	2	5	0	1
12	Ye et al. [22]	2011	Retrospective study	6	375 mg/m^2	2 or 3	1	4	1	1

11 Promising Treatment Alternatives

Table 11.2 The summary of Japanese patients with NPSLE treated with rituximab

	Authors [Ref No.]	Publish year	Article	Patient No.	RTX dose per once	Treated No.	Duration (weeks)	Outcome Complete response	Partial response	Non responder
1	Saito et al. [23]	2003	Case report	1	375 mg/m^2	2	1	1	0	0
2	Tokunaga et al. [24]	2005	Pilot study	2	375 mg/m^2	2	1	2	0	0
3	Tokunaga et al. [25]	2007	Pilot study	10	375 mg/m^2	1 or 2	1	6	4	0
					500 mg/m^2	4	1			
					1000 mg/m^2	2	2			
4	Tanaka et al. [26]	2007	I/II open study	5	1000 mg/m^2	4	1	0	5	0

In the summary of Japanese patients with NPSLE a clinical response was observed in 100% (18/18) of patients, classified as complete response in 50% (9/18) and partial response in 50% (9/18) of patients (Table 11.2). In the summary of NPSLE patients excluding Japanese patients a clinical response was observed in 87% (27/31) of patients, classified as complete response in 81% (25/31) and partial response in 6% (2/31) of patients (Table 11.1). Although different therapeutic regimens of rituximab were used, the high frequencies of the efficacy of rituximab in NPSLE were shown in both studies.

Tokunaga et al. reported a response rate of 100% in a series of ten severe refractory NPSLE patients treated with rituximab [25]. Narvaez et al. also summarized all published data concerning adult patients with refractory NPSLE [20]. A clinical response was observed in 85% (29/34) of patients, classified as complete response in 50% (17/34) and partial response in 35% (12/34) of patients [20]. However, 45% of these patients relapsed after a median of 17 months despite maintenance therapy [20]. Different therapeutic regimens of rituximab were used in these two studies. The most frequently used regimen was 1000 mg doses separated by 15 days. Different dosing schedules appeared to show no difference in response, tolerability, or side effects. In all cases, rituximab was administered together with corticosteroids [20]. A recent case series on 18 pediatric NPSLE patients showed promising effect of rituximab. Thus, the authors divided the benefit into definite (five patients), probable (seven patients), possible (five patients) and no effect (one patient) [27].

Long-term studies regarding rituximab therapy in NPSLE have been rarely performed, thus far. Current data support the use of rituximab as a second-line therapy in patients with severe refractory NPSLE, although additional controlled studies are needed to define the exact place of rituximab in the therapeutic regimen for NPSLE.

11.2.3 Case Report 1: The Efficacy of Rituximab in 33-Year Old Woman with NPSLE

In this section the successful treatment with rituximab of a patient with life-threatening refractory NPSLE (acute confusional state) in 2005 is shown. The clinical course of the patient is shown in Fig. 11.2.

A 33-year old Japanese woman with SLE was hospitalized in November 2005 because of high fever and mutism. Just after admission, her consciousness level deteriorated to coma. An immunological study showed ANA 1:2560 (speckled pattern) and anti-ribonucleoprotein antibodies 156.6 U/ml. Anti-ds-DNA antibodies, anti-Sm, and anti-ribosomal P protein antibodies were negative. Serum complement levels were normal: C3 151 mg/dl, C4 43 mg/dl, and CH50 > 69 U. CSF interleukin-6 levels and CSF anti-neuronal antibody titers were 1.0 U/ml (normal <0.02) and 4.5 U/ml (normal <0.27), respectively. She was initially treated with intravenous of prednisone 100 mg per day followed by intravenous pulses of methylprednisolone (1 g/day for 3 consecutive days). Although coma was improved and CSF interleukin-6 levels were decreased, CSF anti-neuronal antibody titers were rather increased and then she showed hallucination and akinetic mutism. These symptoms did not almost improve despite daily intravenous of prednisone.

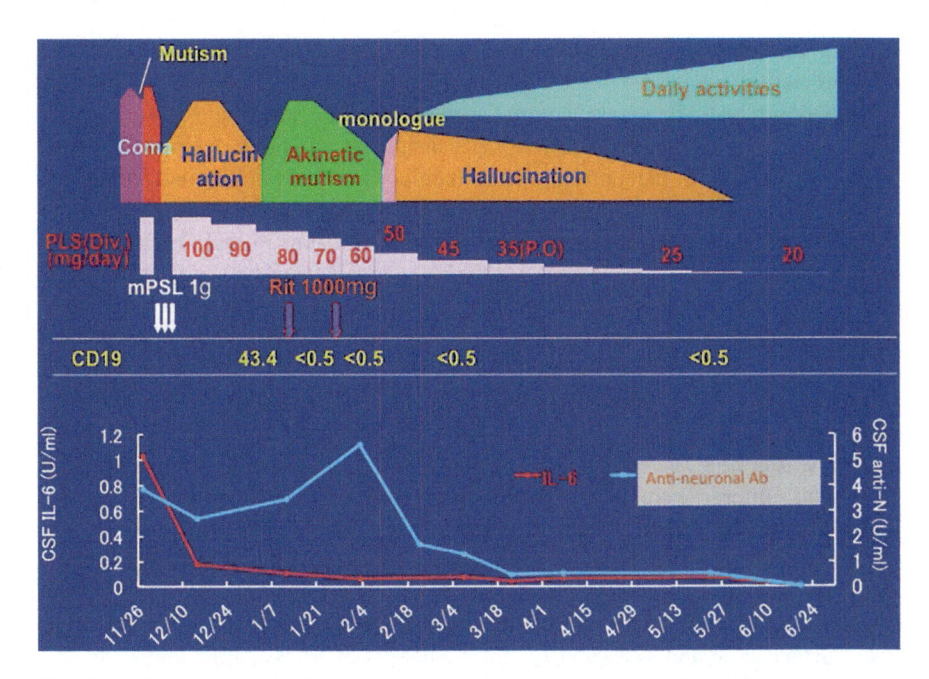

Fig. 11.2 Clinical course and response to treatment with rituximab. Although the patient was treated with high doses of steroid followed by intravenous methylprednisolone (m-PSL pulse, white arrows), CSF anti-neuronal antibodies remained high with modest improvement of psychiatric manifestations. After the addition of rituximab, the improvement was facilitated. (The same patient shown in Fig. 5.1a, b in Chap. 5)

Forty-three days after admission, it was decided to add rituximab to daily intravenous of prednisone in order to decrease CSF anti-neuronal antibody titers. Rituximab was administered at a dose of 1 g separated by a 2-week interval. The numbers of peripheral CD19+ B cells and CSF anti-neuronal antibody titers dramatically decreased after the first and second dose of rituximab. Although she showed monologue after the second dose of rituximab, hallucination and akinetic mutism gradually decreased and her daily activities conversely increased. It appears that addition of rituximab facilitated the decrease of neuron-reactive autoantibodies in CSF, resulting in the amelioration of psychiatric manifestations.

11.2.4 Case Report 2: The Efficacy of Rituximab in 38-Year Old Woman with NPSLE

Narváez et al. reported the successful treatment with rituximab of life-threatening refractory NPSLE in 2011 [20]. The report of a 38-year-old Hispanic woman is shown in this section.

A 38-year-old Hispanic woman, who had been diagnosed with SLE on the basis of arthritis, oral ulcers, leukolymphopenia, and positive antinuclear antibodies (ANA) and anti-dsDNA antibodies, was treated with low doses of prednisone

(10 mg daily), nonsteroidal anti-inflammatory drugs, and hydroxychloroquine (200 mg daily).

One month later she was transferred to the hospital because 3 episodes of generalized tonic-clonic seizures with a 2-week history of cognitive dysfunction manifested by impairments in mental activities (memory, abstract thinking and judgment), which had evolved into disorientation, persecutory delusion, and visual and auditory hallucinations. On admission, she did not show focal neurologic symptoms. An immunological study showed ANA 1:320 (homogeneous pattern) and anti-ds-DNA 459 kint.u./L (normal 14.9). Anti-Sm, anti-ribonucleoprotin antibodies, anticardiolipin antibodies, lupus anticoagulant, and anti-beta-2-glycoprotein 1 antibodies were negative. Serum complement levels were reduced: C3 394 mg/L (normal range: 750–1400), C4 26.6 mg/L (normal: 100–340), and CH50 4 UH (normal: 51–136).

CSF was normal (no cells, total protein 0.25 g/L, and glucose 3.4 mmol/L). CSF and blood cultures were sterile. Brain magnetic resonance imaging and cerebral magnetic resonance angiography were normal.

Despite the treatment with anticonvulsants (phenytoin and valproate), prednisone 1 mg/kg per day with intravenous pulses of methylprednisolone (1 g/day for 3 consecutive days), followed by intravenous-cyclophosphamide pulse (500 mg/m^2 of body surface area) and 3 sessions of plasmapheresis at 20 days after admission, she showed modest improvement. Then, addition of rituximab to the intravenous-cyclophosphamide was decided. Rituximab was administered at a dose of 1 g separated by a 2-week interval (days 1 and 15). The patient responded within a few weeks of the first dose of rituximab, with disappearance of seizures and progressive resolution of the psychiatric symptoms [20]. It is thus suggested that some autoantibodies (possible neuron-reactive) might play a pivotal role in the pathogenesis of this patient like case 1.

11.2.5 Mechanism of Action of Rituximab

The mechanism through which rituximab acts on SLE remains unclear. It substantially reduces levels of CD20+ B cells in human peripheral blood within days to weeks. This effect may be sustained for up to 6 months [24, 28] and subsequent immune reconstitution improved peripheral B-cell abnormalities, including lymphopenia and expansion of autoreactive cells [29].

This suggests that treatment with rituximab might alter the pathology of the disease once B-cell repletion occurs. However, it does not appear to work solely via the diminution of autoantibodies. Mechanisms of action that have been proposed in the literature include complement-mediated cellular lysis, B-cell-triggered apoptosis, and antibody-dependent cellular toxicity [30]. In addition, it has been found that SLE patients receiving this biological agent show a reduced expression of the costimulatory molecules CD40 and CD80 on B cells, and that of CD40L, CD69, and inducible costimulator on CD4 T cells. These findings suggest that rituximab modulates the interaction of activated B and T cells through affecting the expression of costimulatory molecules [30].

The mechanism which rapid recovery of neuropsychiatric symptoms appears in patients with NPSLE who was treated with rituximab remains still unclear. A few mechanisms might be suggested as follows:

1. The depletion of B cells by rituximab cause the decrease of B cell entering the CNS through the blood-brain barrier (BBB).
2. The rapid decrease of autoantibody production after the depletion of B cells by rituximab, resulting the decrease of neuropathic autoantibodies and immune complexes entering the CNS through the BBB and that of complement activation by immune complexes in the BBB and the CNS.
3. The rapid recovery of the damaged BBB, which triggers inflammatory conditions, might block the infiltration of cytokines, chemokines and immune cells such as neutrophils in the circulation into the CNS.

11.2.6 Rituximab as Potential Future Therapies

The evidence for the effectiveness of rituximab as induction therapy in NPSLE is based solely on several case reports and noncontrolled trials. Although it is not yet possible to make definite recommendations, the global analysis of these cases supports the off-label use of rituximab as a second-line therapy in patients with severe refractory NPSLE. The safety profile of B cell depletion therapy is favorable, although ongoing vigilance for adverse reactions is required.

The high rate of efficacy found may be partially explained by the fact that most reports include cases with a favorable response, whereas cases without such a response are not often reported. RCTs are clearly needed to confirm the open-label data and to establish the correct dose, the length of therapy and the appropriate use of concomitant medications. However, as noted above, it must be remembered that inadequately designed trials may yield confounding results. Therefore, when designing future studies, it is important to include patients with high disease activity of NPSLE or those with refractory NPSLE to conventional therapy. Alternatively, the effect of rituximab should be analyzed in comparison to conventional immunosuppressive treatment.

11.3 Potential Future Therapies Other than Rituximab

11.3.1 Anti-CD22 Antibodies (Epratuzumab)

Epratuzumab, a humanized monoclonal antibody that targets CD22 on B cells and results in modulation of B-cell function and migration, has also been studied in SLE patients. Although the EMBLEM and ALLEVIATE trials showed promising results, the EMBODY I and EMBODY II phase III clinical studies for epratuzumab in SLE did not meet their primary clinical efficacy endpoints [31, 32]. However, epratuzumab therapy in NPSLE has not been reported.

11.3.2 B Cell Stimulator Targets

Both B lymphocyte stimulator (BLyS), which is also known as B-cell activating factor (BAFF), and a proliferation-inducing ligand (APRIL), which is also known as tumor necrosis factor ligand superfamily member, were shown to be elevated in the CSF of SLE patients. Furthermore, APRIL was increased in the CSF of NPSLE patients compared with SLE patients without neuropsychiatric symptoms and other neurological diseases. It has been suggested that BLyS and APRIL are produced locally in the astrocytes of the brain and that they may play a role in NPSLE etiology. Hence, antagonists of these cytokines could have beneficial effect in these patients [33, 34].

11.3.2.1 Anti-BLyS Antibodies (Belimumab, Tabalumab and Blisibimod)

Belimumab, a humanized monoclonal antibody targeted against B lymphocyte stimulator (BLyS), is now licensed in the United States of America, Europe and Japan for the management of SLE. The BLISS trials were neither designed nor powered to definitively demonstrate the efficacy of belimumab in specific organ systems.

Blisibimod is a selective antagonist of the BLyS, which was initially developed as a treatment for SLE. Blisibimod is currently being tested in a Phase III study, CHABLIS-SC1, for SLE. A phase III trial of tabalumab, an anti-BLyS human monoclonal antibody, against SLE was terminated early as the study failed to meet its primary endpoint. However, blisibimod or tabalumab therapy in NPSLE has not been reported.

11.3.2.2 BLyS and APRIL Receptor (Atacicept)

Atacicept, a humanized fusion protein that binds BLyS and APRIL has also been tested in SLE patients [35]. To date, results in SLE patients are promising and further studies with these therapies are awaited; however, patients with severe NPSLE were excluded from all these trials, which will limit any conclusion in this respect [36].

11.3.3 Cytokine and Chemokine Targets

Several studies have confirmed the intrathecal presence of higher levels of cytokines and chemokines in NPSLE. High levels of IL-6, IL-8, IP-10, MCP-1, G-CSF, tumor necrosis factor (TNF)-α, and IFN-γ in the CSF of NPSLE patients are reported [37].

The overproduction of these cytokines is thought to play a role in the pathogenesis and severity of NPSLE, and they have been proposed as candidate targets for future treatment [33, 38–40].

Although not confirmed in all studies, IFN-α has been also reported to be one of the inflammatory mediators related to NPSLE pathogenesis. Type I IFNs are found in glia and neurons. Among their functions, IFNs induce other inflammatory mediators such as IL-6, alter brain neurotransmitters such as serotonin, and generate brain toxic metabolites. Subsequently, IFN-α has been hypothesized as a potential target in NPSLE [41–43].

11.3.3.1 Anti-IL-6 Receptor Antibodies (Tocilizumab)

A phase I trial with tocilizumab, anti-IL-6 receptor antibodies, has shown acceptable results and more studies are awaited [44]. Although antibodies are administered intravenously and may have a therapeutic effect on the brain, taking into account the BBB disruption in NPSLE, tocilizumab may require transport across the BBB using an endogenous BBB peptide receptor transporter [45].

11.3.3.2 Anti-IFN-α Antibodies (Sifalimumab and Rontalizumab)

IFN-α is considered one of the most promising therapeutic targets in SLE and NPSLE. Sifalimumab, a human anti-IFN-α monoclonal antibody, and rontalizumab, a humanized monoclonal antibody, have shown promising results in reducing SLE disease activity across multiple clinical measures [46].

A common characteristic of these new therapies is that the impact on disease activity seems promising but must still be assessed in phase III trials. However, since in most of these trials NPSLE was an exclusion criterion, the potential to treat NPSLE will remain unknown [36].

11.3.4 Complement Targets: Anti-Terminal Complment Component C5a And C5b-9 Antibodies (Eculizumab)

Complement component C5 has been reported to play a role in the maintenance of the BBB in mice [47]. Selective inhibition of C5aR alleviated NPSLE [48]. Also, inhibition of the classical and alternative complement cascade with the complement inhibitor Crry was demonstrated to alleviate experimental NPSLE in mice [49]. Furthermore, complement plays a role in microvascular injury. Mice deficient in C3 and C5 components are resistant to enhanced thrombosis and endothelial cell activation induced by anti-phospholipid antibodies, indicating the important role of alternative pathway complement activation on anti-phospholipid antibody-mediated thrombogenesis [50, 51].

Based on the above information, eculizumab, a humanized monoclonal antibody blocking the generation of terminal complement components C5a and C5b-9, may be a potential drug to be used in NPSLE in the future [52].

The BBB is a network of endothelial cells and pericytes with astrocyte projections that regulates the entry of soluble molecules and cells into the brain parenchyma. It has been proposed that a disruption of the integrity of the BBB may have a potential pathogenic role in NPSLE since this may permit the influx of neuropathic antibodies across the BBB. Many modulators of the integrity of the BBB have been proposed. Among them, anti-endothelial cell antibodies including anti-NR2 antibodies and antiribosomal P protein antibodies, complement components, cytokines and chemokines, and environmental mediators have an essential role [53].

11.4 Summary

Neuropsychiatric symptoms constitute an uncommon and poorly understood event in SLE patients, and pose a diagnostic and therapeutic challenge to physicians. Management of NPSLE patients has not evolved substantially in the last decades and is characterized by the lack of good evidence and the use of empirical therapies to date. It seems reasonable that increased understanding of the pathogenesis of NPSLE and any of its manifestations will promote the possibility of finding targeted therapies and an evidence-based approach to management.

References

1. Unterman A, et al. Neuropsychiatric syndromes in systemic lupus erythematosus: a meta-analysis. Semin Arthritis Rheum. 2011;41:1–11.
2. Bertsias GK, Boumpas DT. Pathogenesis diagnosis and management of neuropsychiatric SLE manifestations. Nat Rev Rheumatol. 2010;6:358–67.
3. Sanna G, et al. Neuropsychiatric involvement in systemic lupus erythematosus: current therapeutic approach. Curr Pharm Des. 2008;14:1261–9.
4. Tieng AT, Peeva E. B- cell-directed therapies in systemic lupus erythematosus. Semin Arthritis Rheum. 2008;38:218–27.
5. Ramos Casals M, et al. Rituximab in systemic lupus erythematosus. A systematic review of off-label use in 188 cases. Lupus. 2009;18:767–76.
6. Murray E, Perry M. Off-label use of riyuximab in systemic lupus erythematosus: a systematic review. Clin Rheumatol. 2010;29:707–16.
7. Glennie MJ, et al. Mechanisms of killing by anti-CD20 monoclonal antibodies. Mol Immunol. 2007;44:3823–37.
8. Merrill JT, et al. Efficacy and safety of rituximab in moderately-to-severely active systemic lupus erythematosus: the randomized, double-blind, phase II/III systemic lupus erythematosus evaluation of rituximab trial. Arthritis Rheum. 2010;62:222–33.
9. Rovin BH, et al. Efficacy and safety of rituximab in patients with active proliferative lupus nephritis: the lupus nephritis assessment with rituximab study. Arthritis Rheum. 2012;64:1215–26.

10. Cobo-Ibanez T, et al. Efficacy and safety of rituximab in the treatment of non-renal systemic lupus erythematosus: a systematic review. Semin Arthritis Rheum. 2014;44:175–85.
11. Weide R, et al. Successful longterm treatment of systemic lupus erythematosus with rituximab maintenance therapy. Lupus. 2003;12:779–82.
12. Gottenberg JE, et al. Tolerance and short term efficacy of rituximab in 43 patients with systemic autoimmune diseases. Ann Rheum Dis. 2005;64:913–20.
13. Leandro MJ, et al. B-cell depletion in the treatment of patients with systemic lupus erythematosus: a longitudinal analysis of 24 patients. Rheumatology. 2005;44:1542–5.
14. Armstrong DJ, et al. SLE-associated transverse myelitis successfully treated with rituximab (anti-CD20 monoclonal antibody). Rheumatol Int. 2006;26:771–2.
15. Birnbaum J, Kerr D. Optic neuritis and recurrent myelitis in a woman with systemic lupus erythematosus. Nat Clin Pract Rheumatol. 2008;4:381–6.
16. Abud-Mendoza C, et al. Treating severe systemic lupus erythematosus with rituximab. An open study. Reumatol Clin. 2009;5:147–52.
17. Lu TY, et al. A retrospective seven-year analysis of the use of B cell depletion therapy in systemic lupus erythematosus at university college London hospital: the first fifty patients. Arthritis Rheum. 2009;61:482–7.
18. Nasir S, et al. Nineteen episodes of recurrent myelitis in a woman with neuromyelitis optica and systemic lupus erythematosus. Arch Neurol. 2009;66:1160–3.
19. Espinosa G, et al. Transverse myelitis affecting more than 4 spinal segments associated with systemic lupus erythematosus: clinical, immunological, and radiological characteristics of 22 patients. Semin Arthritis Rheum. 2010;39:246–56.
20. Narvaez J, et al. Rituximab therapy in refractory neuropsychiatric lupus: current clinical evidence. Semin Arthritis Rheum. 2011;41:364–72.
21. Pinto LF, et al. Rituximab induces a rapid and sustained remission in Colombian patients with severe and refractory systemic lupus erythematosus. Lupus. 2011;20:1219–26.
22. Ye Y, et al. Rituximab in the treatment of severe lupus myelopathy. Clin Rheumatol. 2011;30:981–6.
23. Saito K, et al. Successful treatment with anti-CD20 monoclonal antibody (rituximab) of life-threatening refractory systemic lupus erythematosus with renal and central nervous system involvement. Lupus. 2003;12:798–800.
24. Tokunaga M, et al. Down-regulation of CD40 and CD80 on B cells in patients with life-threatening systemic lupus erythematosus after successful treatment with rituximab. Rheumatology. 2005;44:176–82.
25. Tokunaga M, et al. Efficacy of rituximab (anti-CD20) for refractory systemic lupus erythematosus involving the central nervous system. Ann Rheum Dis. 2007;66:470–5.
26. Tanaka Y, et al. A multicenter phase I/II trial of rituximab for refractory systemic lupus erythematosus. Mod Rheumatol. 2007;17:191–7.
27. Dale RC, et al. Utility and safety of rituximab in pediatric autoimmune and inflammatory CNS disease. Neurology. 2014;83:142–50.
28. Sfikakis PP, et al. Remission of proliferative lupus nephritis following B cell depletion therapy is preceded by down-regulation of the T cell costimulatory molecule CD40 ligand: an open-label trial. Arthritis Rheum. 2005;52:501–13.
29. Anolik JH, et al. Rituximab improves peripheral B cell abnormalities in human systemic lupus erythematosus. Arthritis Rheum. 2004;50:3580–90.
30. Driver CB, et al. The B cell in systemic lupus erythaematosus: a rational target for more effective therapy. Ann Rheum Dis. 2008;67:1374–81.
31. Harvey PR, Gordon C. B- cell targeted therapies in systemic lupus erythematosus: successes and challenges. BioDrugs. 2013;27:85–95.
32. Study of epratuzumab versus placebo in subjects with moderate to severe general systemic lupus erythematosus (SLE) (EMBODY 2). Available at: https://clinicaltrials.gov/ct2/show/NCT01261793. Accessed 22 Nov 2015.

33. George-Chandy A, et al. Raised intrathecal levels of APRIL and BAFF in patients with systemic lupus erythematosus: relationship to neuropsychiatric symptoms. Arthritis Res Ther. 2008;10:R97.
34. Hopia L, et al. Cerebrospinal fluid levels of a proliferation-inducing ligand (APRIL) are increased in patients with neuropsychiatric systemic lupus erythematosus. Scand J Rheumatol. 2011;40:363–72.
35. Manzi S, et al. Effects of belimumab, a B lymphocyte stimulator-specific inhibitor, on disease activity across multiple organ domains in patients with systemic lupus erythematosus: combined results from two phase III trials. Ann Rheum Dis. 2012;71:1833–8.
36. Ding HJ, Gordon C. New biologic therapy for systemic lupus erythematosus. Curr Opin Pharmacol. 2013;13:405–12.
37. Yoshio T, et al. IL-6, IL-8, IP-10, MCP-1 and C-CSF are significantly increased in cerebrospinal fluid but not in sera of patients with central neuropsychiatric lupus erythematosus. Lupus. 2016;25:997–1003.
38. Fragoso-Loyo H, et al. Interleukin-6 and chemokines in the neuropsychiatric manifestations of systemic lupus erythematosus. Arthritis Rheum. 2007;56:1242–50.
39. Svenungsson E, et al. Increased levels of proinflammatory cytokines and nitric oxide metabolites in neuropsychiatric lupus erythematosus. Ann Rheum Dis. 2001;60:372–9.
40. Trysberg E, et al. Intrathecal cytokines in systemic lupus erythematosus with central nervous system involvement. Lupus. 2000;9:498–503.
41. Dafny N, Yang PB. Interferon and the central nervous system. Eur J Pharmacol. 2005;523:1–15.
42. Fragoso-Loyo H, et al. Utility of interferon-alpha as a biomarker in central neuropsychiatric involvement in systemic lupus erythematosus. J Rheumatol. 2012;39:504–9.
43. Santer DM, et al. Potent induction of IFN-alpha and chemokines by autoantibodies in the cerebrospinal fluid of patients with neuropsychiatric lupus. J Immunol. 2009;182:1192–201.
44. Illei GG, et al. Tocilizumab in systemic lupus erythematosus: data on safety, preliminary efficacy, and impact on circulating plasma cells from an open-label phase I dosage-escalation study. Arthritis Rheum. 2010;62:542–52.
45. Pardridge WM. Targeted delivery of protein and gene medicines through the blood-brain barrier. Clin Pharmacol Ther. 2015;97:347–61.
46. Mathian A, et al. Targeting interferons in systemic lupus erythematosus: current and future prospects. Drugs. 2015;75:835–46.
47. Jacob A, et al. C5a alters blood-brain barrier integrity in experimental lupus. FASEB J. 2010;24:1682–8.
48. Jacob A, et al. Inhibition of C5a receptor alleviates experimental CNS lupus. J Neuroimmunol. 2010;221:46–52.
49. Alexander JJ, et al. Administration of the soluble complement inhibitor, Crry-Ig, reduces inflammation and aquaporin 4 expression in lupus cerebritis. Biochim Biophys Acta. 2003;1639:169–76.
50. Cavazzana I, et al. Complement activation in anti-phospholipid syndrome: a clue for an inflammatory process? J Autoimmun. 2007;28:160–4.
51. Thurman JM, et al. A novel inhibitor of the alternative complement pathway prevents antiphospholipid antibody-induced pregnancy loss in mice. Mol Immunol. 2005;42:87–97.
52. Barilla-Labarca ML, et al. Targeting the complement system in systemic lupus erythematosus and other diseases. Clin Immunol. 2013;148:313–21.
53. Diamond B, et al. Losing your nerves? Maybe it's the antibodies. Nat Rev Immunol. 2009;9:449–56.

Chapter 12
Prognosis of Neuropsychiatric Systemic Lupus Erythematosus

Shinsuke Yasuda

Abstract It is still a matter of debate whether mortality rate is higher in SLE patients with neuropsychiatric (NP) symptoms compared to those without them. Probably this situation is based on the difference in the character of each cohort and also on the vagueness of the definition/diagnosis of NPSLE, even after 1999. Although NPSLE has been repeatedly detected as one of the predictive factors for poor prognosis, NPSLE itself is not a common direct cause of death. It is more likely that patients with NPSLE tend to suffer from more recalcitrant lupus disease activities requiring intense immunosuppressive treatment or sometimes too severe to control. However, more and more advanced brain imaging, understanding on neuroscience and pathophysiology of NPSLE, and novel targeted-therapies are emerging. Thus we would expect better prognosis, cognitive/psychological functions and qualities of life for patients with NPSLE in near future.

Keywords Prognosis · Mortality · Irreversible damage · Cerebrovascular disease · Cognitive dysfunction

12.1 Introduction

Prevalence of NPSLE would vary among ethnicity, among reports and between genders. Moreover, the results are affected by how (prospective or retrospective), when and where the cohort or case series were collected and analyzed, and by how NPSLE was diagnosed. Although nomenclature and case definition of NPSLE was proposed by the American College of Rheumatology (ACR) in 1999 [1], it is yet to

S. Yasuda (✉)
Department of Rheumatology, Endocrinology and Nephrology, Faculty of Medicine and Graduate School of Medicine, Hokkaido University, Sapporo, Japan
e-mail: syasuda@med.hokudai.ac.jp

establish the diagnostic criteria. Therefore, it is rather natural that reported prevalence of NP symptoms in lupus patients shows wide variations.

Qualified meta-analyses that include large population of the patients might reinforce and help our knowledge on the clinical aspects of NPSLE. For example, according to a meta-analysis concerning the epidemiology of NPSLE [2], the prevalence of NPSLE in patients with SLE was more than 40% in prospective studies but was less than 20% in retrospective ones. Prevalence of each manifestation in NPSLE also differed significantly according to study design. These discrepancies might come from vagueness in the diagnostic process for NP symptoms, especially for non-specific NP symptoms such as "headache" or "mood disorder". One can easily imagine how it is difficult to extract "lupus headache" from SLE non-related headache referring to clinical charts, especially in a retrospective analysis. We always need to take such limitations into account when discussing epidemiology, treatment efficacy, and prognosis of NPSLE.

Definition for the prognosis of NPSLE is also vague. How can we define/evaluate the prognosis of NPSLE? Does complication of NPSLE increase mortality of the patients? Which NPSLE symptoms tend to recover completely and which ones tend to be irreversible? In this chapter, I would like to explain the difficult situation we are standing now, and then try to answer above questions looking at the real-world clinical data concerning the prognosis of NPSLE.

12.2 Mortality in Patients with NPSLE

It is not clear whether mortality rate is higher in patients with NPSLE compared to those without. According to relatively old-days' reports before publication of ACR nomenclature and case definition for NPSLE, increment of mortality rate in patients with NPSLE was controversial. In one Canadian study, most CNS events were self-limited, reversible and not associated with poor prognosis unless accompanied by multisystem disease activity [3]. In another study from Greece, 32 hospitalized NPSLE patients were followed-up for 2-years, mostly concerning NP deficits with brain MRI, resulting in no death in these patients [4]. In 1980s, a multicenter study including more than a thousand SLE patients was carried out to clarify the mortality and disease characteristics [5]. Rate of survival decreased in patients with CNS features of seizures ($P < 0.05$) and organic psychosis ($P < 0.05$). In a Swedish prospective study, SLE patients diagnosed during 1981–1995 were recruited and followed up till 1998. NP manifestations developed in 38% of the patients, but mortality rate was not increased compared with SLE patients without NP symptoms (NPSLE: 6 deaths in 44 versus non-NPSLE: 13 in 73, direct causes for death not provided) [6].

When it comes down to more recent studies, larger sized cohort studies based on ACR nomenclature become available. According to a prospective analysis of an international disease inception cohort of 1206 SLE patients, NP events attributed to SLE occurred in 17.7% to 30.6% during mean follow-up period of 1.9 years. In this analysis, headache was treated as non-SLE NP events. Among SLE NP events, "seizures" were the most frequent, followed by "mood disorder" and "cerebrovascular disease". There

12 Prognosis of Neuropsychiatric Systemic Lupus Erythematosus

Table 12.1 Cause/number of deaths in patients with NPSLE

Cohort (reference) (Patients number)	International [7] (SLE, 1206)	Dutch [8] (NPSLE, 169)	Japan [9] (NPSLE, 53)	Japan [10] (NPSLE, 79)
NP symptoms	4	5	1	1
Intracranial hemorrhage	(2)	(1)		
Stroke	(1)	(1)		
Seizure	(1)	(1)		
Aseptic meningitis		(1)		
Vasculitis/cerebritis		(1)	(1)	
Suicide				(1)
Other SLE disease activity	NA	4	4	2
Infection	NA	7	3	4
Malignancy	NA	4	NA	0
Cardiovascular	NA	0	1	0
Other	NA	1	0	1
Unknown	NA	11	0	2
Sum	18	32	9	10
SMR	NA	9.5	NA	NA
Risk factor		ACS	MRI	
Observation (mean, years)	1.9	6 (median)	6	4.8

were 18 (1.5%) deaths, in 4 cases of which the primary cause of death was attributed to NP events such as intracranial hemorrhage, stroke and seizures [7] (Table 12.1).

In a recent single-center study conducted in Netherland took advantage of the civic registries [8]. All suspected NPSLE patients were reevaluated based on ACR nomenclature and case definition, then followed up using national civic registries. Thirty-two (19%) of the 169 NPSLE patients died within a median follow-up period of 6 years, bringing the standardized mortality ratio (SMR) as high as 9.5. Causes of death in these patients were as follows: infection in 7, NPSLE in 5 (brainstem hemorrhage, epileptic state, hemorrhagic infarction, aseptic meningitis, vasculitis of brain arteries), other SLE-related disease activity in 4, cancer in 4, pneumothorax in 1 and unknown causes in 11. Multivariate analysis revealed that "acute confusional state" was the highest risk with hazard ratio of 3.4, and that older age at onset of NPSLE slightly increased the risk of mortality. Clinical study calculating SMR is limited in NPSLE, but it seems that SMRs were somewhat higher in NPSLE compared with those reported in whole SLE or lupus nephritis. For example, in our biopsy-proven lupus nephritis patients (N = 186), SMR was 3.59 [11]. Other reported SMR in Chinese patients with lupus nephritis ranges from 5.9 to 9.0 [12, 13].

Arinuma et al. [9] analyzed MRI findings and outcome of 53 patients with diffuse NPSLE, including 37 with "acute confusional state". During observation periods of 73 months at mean, 9 (17%) patients died: 3 from pneumonia, one from rupture of aortic aneurysm, and the remaining 5 with active SLE including pulmonary hypertension, catastrophic APS, thrombotic thrombocytopenic purpura, pneumatosis intestinalis, and cerebritis. In this setting of patients, those with abnormal findings in brain MRI had significantly poorer overall survival, comparing with those without MRI abnormality [9]. In our retrospective analysis on 79 SLE patients who developed NP symptoms before or after initiation of treatment, 70 were diagnosed as having NPSLE. Among them, 10 (14%) patients died, including 4 due to infection, 2 to sudden death of unknown reason, one alveolar hemorrhage, one acute pancreatitis, one hemophagocytic syndrome, and one patient committed suicide [10].

In a large international Latin American cohort, effect of antimalarials on patients' survival was investigated [14]. In their multivariate analysis, neurologic disorder at diagnosis was defined as a risk for mortality with a hazard ratio of 1.73, although ACR 1999 nomenclature was not mentioned.

According to recent cohorts/case series, mortality rate in SLE patients with NP symptoms seems to be relatively high, although NP symptoms are not direct cause of death in many cases. Rather, patients with NP symptoms tend to suffer from more severe multi-organ involvement and require intensive immunosuppressive treatment, resulting in opportunistic infection and/or uncontrollable lupus disease activity.

12.3 Evaluation of NP Symptoms and Irreversible NP Damages in SLE

Sustaining or irreversible NP symptoms impact patients' health-related quality of life (HRQoL), thus confounding unmet needs in clinical practice [15, 16]. Irreversible damages should be accounted for as a matter of discussion when we consider the prognosis of NPSLE. However, what is the definition of irreversible damages in NP systems? Neurological damage due to "cerebrovascular disease" is the most obvious and objective event included in this category. Then, how about psychiatric symptoms such as "cognitive dysfunction" or "psychosis"? These symptoms are irreversible in some cases, while in other cases long-standing and irreversible-looking "cognitive dysfunction" or "mood disorder" represented by depression gradually improve or even disappear. In addition, these psychiatric symptoms are sometimes difficult to evaluate, especially for rheumatologists who are not specialized for psychology or neurology in many cases.

Laboratory tests recommended by ACR include the assessment of lupus disease activity, antiphospholipid antibodies, and in limited circumstances, anti-ribosomal P antibodies. Radiologic tests include the use of computed tomography, MRI, angiography, electrocardiography, echocardiography and duplex ultrasound [1]. As mentioned in the same literature [1], "focal neurologic syndromes appear to be diagnosed with little disagreement". In contrast, more problematic domain is psychiatric disorders, cognitive deficits, and acute confusional states. The Systemic Lupus International Collaborating Clinics/American College of Rheumatology (SLICC/ACR) damage index (SDI) comprises the following 4 items in the NP category: "cognitive impairment", "seizures requiring therapy for 6 months", "cerebrovascular accident ever", "cranial or peripheral neuropathy", and "transverse myelitis" [17].

In general, MRI abnormality is associated with cerebrovascular disease, but changes in white matter microstructure were proposed as suggestive for inflammatory NPSLE [18]. Magnetizing transfer imaging (MTI) is a quantitative MRI technique useful in the detection of abnormalities in brain tissue that looks normal on conventional MRI. Among MTI parameters, the histogram peak height (HPH) is the most informative parameter in NPSLE without explanatory MRI findings. Moreover, parallel improvement of clinical status and cerebral changes in white matter using quantitative MRI of the patients after immunosuppressive treatment have reported. Such sophisticated MRI-derived parameter is emerging and opening a new horizon for the evaluation of non-ischemic NPSLE.

Evaluation of psychiatric syndromes including "cognitive dysfunction", "mood disorder", "anxiety disorder" and "psychosis" largely depends on experienced psychiatrists. However, many cases with such psychiatric syndrome may be based on inflammatory NPSLE. Therefore, classification/evaluation of NPSLE should ideally become more and more multidisciplinary in order to ensure preciseness in diagnosis/evaluation as follows.

In the Leiden cohort of 100 SLE patients, NP events were prospectively and intensively evaluated [19]. NP events were diagnosed by multidisciplinary evaluation and divided into non-NPSLE, inflammatory NPSLE or ischemic NPSLE. Their multidisciplinary assessment includes SLEDI-2 K, SDI (SLICC/ACR Damage Index), blood and urine laboratory tests, neuropsychological evaluation and brain MRI routinely. Spinal fluid examination and spinal MRI were performed if necessary. HRQoL was also evaluated using Short Form 36 (SF-36) health survey questionnaire. Detailed information is available in the report [19], in which all patients were evaluated by rheumatologists, internists, neurologists, psychiatrists and neuropsychologists at inclusion, then were re-evaluated by all of them plus radiologists. Consensus meeting were held for decision-making for each patient. MTI, resting state functional MRI, Hospital Anxiety and Depression Scale (HADS), Dissociation Experience Scale (DES), Neuropsychiatric Inventory (NPI) were performed at inclusion. Using such extensive collaborative approach requiring one-day admission, patients suspected for NPSLE were classified into "No NPSLE", "Primary NPSLE" comprising ischemic, inflammatory and undefined NPSLE and "Secondary NPSLE". At inclusion, ischemic NPSLE was associated with anticardiolipin antibodies and secondary NPSLE was with history of renal disease and corticosteroid. Symptoms were mild in secondary NPSLE. These descriptions seem quite natural and understandable for many rheumatologists, but inflammatory NPSLE did not have specific features, probably because this category harbors too many aspects of NPSLE. Most of all, it is admirable that such an extensive evaluation has been done in a relatively large cohort in a real-world clinical situation. Certainly this is the way we should go for the assessment and treatment of patients with NPSLE, but there would be many real-world situations in many countries/districts to cope with, such as cost and manpower to conduct this kind of multidisciplinary examination.

12.4 Prognosis of Overall/Specific Manifestations of NPSLE

According to the above-mentioned follow-up study of relatively small numbers of hospitalized patients with NPSLE, NP deficits were mostly improved or stabilized, with deterioration only in 12% [4]. In the Swedish cohort of 117 SLE patients [6], organ damage evaluated by SDI was higher in NPSLE patients compared with non-NPSLE patients and with the general population. Working incapability was also higher in patients with NPSLE compared with other groups.

More recently, by the European international cohort [7], the rate of resolution of NP events attributed to SLE was significantly better than those due to non-SLE causes (52–55% in SLE causes versus 36–38% in non-SLE causes, according to different statistical models). Higher resolution rate was observed in focal NP events compared with diffuse NP events (53% versus 38%). Resolution was more frequently observed if the NP events occurred in earlier phase of the onset of SLE. Similarly, mental (MCS) and physical (PCS) component summary score of the SF-36 was lower in patients with NP events than those without NP events almost constantly. Among patients with NP events, MCS score in SLE-related NP events seems to be reduced overtime, while that in non-SLE NP events does not.

In the follow-up study of the Leiden cohort now including 131 SLE patients, NP events were prospectively and intensively evaluated [16]. A total of 232 NP events were diagnosed by multidisciplinary evaluation. After re-assessment, 19% of all NP events resolved, 33% improved, 35% unchanged and 14% worsened. Notably, two-third of NPSLE events improved or resolved, whereas only one-third of non-NPSLE events did, suggesting the efficacy of immunosuppressive treatment in the NPSLE group. Similarly, inflammatory NPSLE improved in more than half of the patients, whereas ischemic NPSLE ameliorated only in about 16% of the patients.

At a glance, the results of above two studies seem to conflict each other [7, 16], but there is a difference in the categorization in the NPSLE group between these two studies: "diffuse and focal" versus "inflammatory and ischemic". For example, seizure is classified into "focal" in the international cohort, but mostly into "inflammatory" in the Leiden cohort if the seizure was not clearly due to cerebral infarction. Because ACR nomenclature and case definition is not based on pathophysiology, one category of NPSLE symptom may comprise different disease status, resulting in discordant recovery rates or prognosis. We need to take such situations into account when considering/evaluating the prognosis of each NP manifestation in different studies. Here we would like to start with more objectively evaluable part, cerebrovascular disease.

12.4.1 Cerebrovascular Disease

High disease activity and other risk factors such as persistently positive moderate-to-high titers of anti-phospholipid antibodies (aPLs) and common risk factors are associated with cerebrovascular disease in SLE patients. The prevalence of cerebral infarction in SLE or anti-phospholipid syndrome (APS) differs among populations or reports. In our retrospective cohort of consecutive 141 APS patients in Japan, the rate of cerebral infarction was 61% [20], whereas in the international European cohort of a thousand APS patients cerebral infarction was found only in 20% of the patients [21]. Such difference may be due to the difference in the detection methods, in genetic predisposition for venous thrombosis, or in the prevalence of non-APS risk factors represented by hypertension.

How about mortality related with cerebrovascular diseases in SLE? According to a Canadian cohort study of cerebrovascular mortality in SLE comprising 2688 patients with average follow-up of over 9 years [22], 10 patients died due to "cerebrovascular disease". Deaths due to cerebral infarctions appeared to be less common than those due to hemorrhages and others.

In patients with autoimmune diseases, risk of thrombosis including cerebral infarction was assessed using laboratory test-based scoring that includes several aPLs [23]. In this study, we defined aPL-score (or Otomo score) according to the titers of aPLs including lupus anticoagulant (LA) tests, anticardiolipin antibodies (aCL, IgG and IgM), anti-β_2-glycoprotein I antibodies (aβ_2GPI, IgG and IgM) and phosphatidylserine-dependent antiprothrombin antibodies (aPS/PT, IgG and IgM). We found that patients with aPL-score equal or more than 30 are significantly more likely to develop thrombosis compared with those with aPL-score lower than 30 (Table 12.2). Similar scoring system but including conventional clinical risk factors was proposed next year from European group [24]. This scoring system named Global Anti-phospholipid Syndrome Score (GAPSS) includes aPLs and classical thrombotic risk factors such as hypertension, hyperlipidemia and diabetes (Table 12.2). This cross-sectional study included consecutive SLE patients. Patients with higher GAPSS experienced thrombosis and/or pregnancy loss. When the GAPSS cutoff was set at 10, area under the curve (AUC) of the receiver operating characteristic (ROC) curve was the highest for the presence of a history of thrombosis and/or pregnancy loss. Recently, we reviewed these two scoring system and concluded that both aPL-score and GAPSS reach a certain degree of accuracy in identifying high-risk APS patients, especially for thrombosis [25]. However, it is uncertain whether these scores are similarly useful for the prediction of cerebral infarction in patients with SLE.

As mentioned above and as one can easily imagine, ischemic NPSLE is more likely to be irreversible and does not respond to immunosuppressive therapy [16].

12.4.2 Cognitive Dysfunction

Cognitive impairment can serve as an indicator of overall brain health, which can be affected by a number of factors including other NP syndromes [26]. Cognitive dysfunction is one of the common NP events in patients with SLE, but there is no specific pattern attributed to SLE. Therefore, it is rather understandable that trial for the screening of "cognitive dysfunction" in SLE patients using Cognitive Symptom

Table 12.2 Two antiphospholipid scores to predict the risk of thrombosis in patients with autoimmune diseases or SLE

Antiphospholipid-score (Otomo score)		
aPL tests	aPL tests	aPL score
aPTT mixing (sec)	>49	5
aPTT confirmation (ratio)	>1.3	2
	>1.1	1
KCT mixing (sec)	>29	8
dRVVT mixing (sec)	>45	4
dRVVT confirmation (ratio)	>1.3	2
	>1.1	1
IgG aCL high (GPL)	>30	20
IgG aCL low/medium (GPL)	>18.5	4
IgM aCL (MPL)	>7	2
IgG aβ$_2$GPI high (units)	>15	20
IgG aβ$_2$GPI low/medium (units)	>2.2	6
IgM aβ$_2$GPI (units)	>6	1
IgG aPS/PT high (units)	>10	20
IgG aPS/PT low/medium (units)	>2	13
IgM aPS/PT (units)	>9.2	8
Global Anti-phospholipid Syndrome Score (GAPSS)		
aPL or risks	GAPSS	
aCL (IgG and/or IgM)	5	
aβ$_2$GPI (IgG and/or IgM)	4	
aPS/PT (IgG and/or IgM)	3	
LA	4	
Hyperlipidemia	3	
Arterial hypertension	1	

aβ$_2$GPI anti-β$_2$-glycoprotein I antibodies, *aCL* anticardiolipin antibodies, *aPS/PT* phosphatidylserine dependent antiprothrombin antibodies, *KCT* Kaolin clotting time, *LA* lupus anticoagulant

Inventory (CSI) self-report questionnaire turned out to be unreliable [27]. The definition/diagnosis of "cognitive dysfunction" differs among reports, making the prevalence of this NP symptom ranging from 17 to 59% of SLE patients [28]. The diagnostic criteria for cognitive dysfunction require documentation by neurological testing and a decline from a higher former level of functioning [1]. Even short version of neurological test batteries takes approximately 1 h. Nevertheless, it is highly likely that subclinical deficits in cognitive function are more prevalent in patients with SLE.

In a relatively antique, but long-term prospective cohort of 70 SLE patients, 17 out of the evaluated 47 patients (36%) suffered from cognitive impairment during 5-years' follow-up. Cognitive impairment resolved in 9, emerged or fluctuated in 6, and stably impaired in 2 patients [29]. According to more recent prospective cohort of 28 SLE patients with a mean follow-up period of 5-years, majority of the NP variables in standardized neurological tests remained unchanged and minority of the variables improved over time, leading to the conclusion that "cognitive dysfunction" seemed to be a relatively stable feature of CNS involvement in SLE [30].

12.4.3 Acute Confusional State (ACS)

ACS is defined as disturbance of consciousness or level of arousal with reduced ability to focus, maintain, or shift attention, accompanied by cognitive disturbance and/or changes in mood, behavior, or affect [1]. ACS is one of the most severe forms of NPSLE that can progress to coma. The prevalence of ACS in patients with SLE again differs among reports, ranging from less than 10% [31–33] to as many as half of the patients [9, 34]. In our retrospective cohort, 15 patients (18%) were diagnosed as ACS in 79 patients with NPSLE [10], 2 of whom died during 4.8 years' mean observation period. Abe et al. [35] exclusively evaluated the prognosis of 36 ACS patients and found 8 deaths in 18 patients with abnormal brain MRI, whereas no deaths was observed in 18 patients with normal MRI, confirming that abnormalities in brain MRI predict poor overall survival. In fact, the presence of ACS increases hazard ratio for mortality to about 3 in the above referred Dutch NPSLE cohort [8]. In the same study, male gender, presence of serum anti-Sm antibodies and high serum IL-6 also affected the survival in patients with ACS [35]. ACS is one of the inflammatory NP symptoms, thus likely to respond to intensive immunosuppressive therapy. However, it should be pointed out that patients with ACS are in many cases complicated with other organ involvements reflecting extremely high disease activity, such as proliferative nephritis, hemophagocytic syndrome, and/or even diffuse alveolar hemorrhage.

12.5 Treatment that Impact Prognosis of NPSLE

Various immunosuppressive treatments have been introduced for a variety of NP symptoms in patients with SLE. For example, high dose corticosteroids, methylprednisolone pulse therapy, intravenous immunoglobulins, plasma exchange, immunosuppressants including cyclophosphamide, azathioprine, mycophenolate mofetil, and biologics represented by rituximab, have shown clinical efficacy mostly in case series or anecdotal clinical reports, although there are a couple of randomized controlled trial (RCT) comparing the efficacy of different treatment strategy or different drugs.

12.5.1 Inflammatory NPSLE

In a controlled clinical trial recruiting severe NPSLE, treatment effectiveness was compared between mPSL pulse therapy and intravenous cyclophosphamide (IVCY) [36]. Thirty-two patients complicated with severe NPSLE including seizures, peripheral neuropathy, optic neuritis, transverse myelitis etc. were included in this study, received 3 g of mPSL pulse, then randomized into mPSL pulse (bimonthly every 4 months) or monthly IVCY (0.75 g/m^2) for 1 year, followed by the same dose of CY but every 3 months for another year. Response to the treatment according to the criteria suggested by Neuwelt et al. [37] was observed in 18 of 19 patients receiving IVCY, whereas 7 out of 13 patients treated with mPSL pulse responded to the treatment. Obviously, mean number of seizures significantly decreased overtime in the IVCY group. Daily doses of corticosteroids at 6 months and 12 months were significantly fewer in the IVCY group compared with mPSL pulse group.

However, adverse effects on ovary, on bone marrow and immune system, hemorrhagic cystitis and cardiotoxicity by high-dose IVCY are notorious. That is why efficacy of low-dose IVCY was examined for NPSLE in another RCT [38]. Sixty patients with primary NPSLE were randomly assigned to low-dose IVCY (200-400 mg, monthly) plus relatively low-dose daily prednisolone or the same dose of prednisolone only. Patients with lupus nephritis, fever, hypoxia, infection and other severe status were excluded. NP manifestations were mainly represented by cognitive dysfunction, optic neuropathy, stroke/transient ischemic attacks, mood disorder. Thus, really severe status of NPSLE does not seem to be included in this study. As a result, clinical improvement, relapse rate, electro-physiological test preferred better for addition of low-dose IVCY.

No other RCTs concerning induction therapy for NPSLE have been found, whereas plasma exchange [39], IVIg [40], and/or rituximab [41, 42] have been used in severe/refractory NPSLE.

For a maintenance treatment, azathioprine has been used for patients with NPSLE, without evidence from RCTs. In a study from 1970s, among patients with CNS or severe renal disease, azathioprine-treated group showed improved long-term survival and fewer hospitalization, compared with the group of patients who did not receive this immunosuppressant [43]. In a real world, other immunosuppressants such as mycophenolate mofetil and calcineurin inhibitors may be prescribed for patients with NPSLE as a maintenance therapy, although there is no study focusing on the effects of these drugs exclusively for NPSLE.

12.5.2 Ischemic NPSLE

Acute-phase treatment of ischemic stroke is not different from that of non-SLE patients. However, window of opportunity for such thrombolysis therapy is relatively narrow. Therefore, secondary prophylaxis for another ischemic attack becomes of importance to prevent progressive deterioration of the cerebral function. Such treatment strategy includes tight controls of risk factors and antiplatelet therapy. In addition, anticoagulation for prolonged period is recommended in those with APS [44, 45]. However, it is still controversial which combination of antiplatelet therapy and anticoagulation therapy is the most effective for the secondary prevention of ischemic stroke in APS patients.

Intensity of anticoagulation is also a matter of debate. Superiority of high-intensity warfarin (target INR 3.1–4.0) over moderate intensity warfarin (target INR 2.0–3.0) for secondary thromboprophylaxis was demonstrated by RCTs, but risk for minor bleeding was increased by the high-intensity warfarin [46, 47]. Then, how about the effect of direct oral anticoagulants (DOACs) on these settings, comparing to warfarin? One study which compared rivaroxaban with standard warfarin treatment reported no thrombotic or major bleeding events, but this study was underpowered to draw any conclusions [48]. Thus, there is not enough evidence yet to support or decline the use of DOAC for the secondary prevention of cerebrovascular events related to APS. There is not enough evidence that shows benefit/harm of warfarin plus antiplatelet agent or dual antiplatelet therapy (DAPT) comparing to a single antiplatelet therapy.

In our cohort of 90 Japanese APS patients with arterial thrombosis (81 with cerebral infarction), patients were treated with single antiplatelet agent, warfarin only (INR 2.0–3.5), warfarin plus single antiplatelet (INR 2.0–3.0), or DAPT for secondary prophylaxis, in a non-randomized way. Retrospective observation for the mean follow-up period of 8-years revealed lower recurrence rate in the DAPT group compared with the warfarin only group (Ohnishi N, Fujieda Y et al., manuscript in preparation). There was no significant difference in the overall survival among all treatment groups.

12.5.3 Corticosteroid-Induced Psychosis or NPSLE?

Corticosteroid (CS)-induced psychosis occurs in less than 10% of the SLE patients, sometimes making differential diagnosis challenging from NP symptoms due to SLE itself [44, 49]. According to our retrospective autoimmune cohort, 36 (24.7%) out of 146 SLE patients treated with CS (prednisolone more than 40 mg/day) developed NP symptom, whereas only 12 (7.4%) out of 162 patients with other autoimmune diseases did. Among these 36 lupus patients, only 9 were diagnosed as having CS-induced psychiatric disorders, the rest 27 patients being diagnosed as NPSLE. Regardless of the diagnosis patients who developed NP symptoms after initiation of CS treatment were categorized as having post-steroid NP manifestations (PSNP-SLE). These patients had better event-free/overall survival compared with SLE patients who already had NP manifestations on admission (*de novo* NPSLE) [10]. When patients diagnosed as steroid-induced psychosis was excluded, statistical significance between these two groups was lost, but the same tendency remained (Fig. 12.1). Such difference may be due to the variation of the symptoms/severity between these two groups of patients and/or to the prompt assessment and treatment given for the in-hospital PSNP-SLE patients.

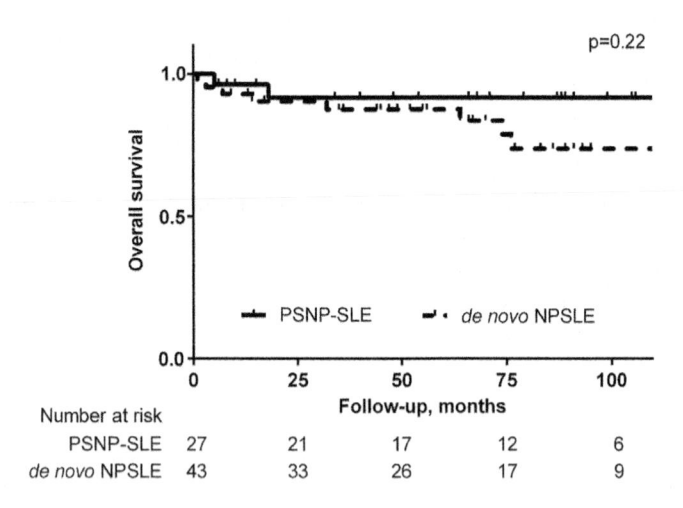

Fig. 12.1 Overall survival of the patients with post-steroid (PS) NPSLE (steroid psychosis excluded) and those with *de novo* NPSLE. Kaplan-Meier overall survival were evaluated. Five-years overall survival rates were 91.7% in post-steroid NPSLE vs. 87.6% in *de novo* NPSLE

12.6 Summary

NP symptoms comprise most severe forms of SLE, which exist solely or concomitantly with other symptoms/organ involvements reflecting the disease activity. Because of the small patient number and the variation of severity of the disease, it is hard to conduct RCTs for NPSLE patients in many cases. In addition, pathophysiology of NPSLE is yet to be clarified in contrast with that of lupus nephritis, partly because of the accessibility to the organ and of the wide variety of symptoms. However, there are growing choices of treatment including targeted antibodies or small molecules approved or under development for SLE, which we might be able to use also as a treatment for NPSLE. More and more progresses have been made in the field of neuroscience, autoimmunity and genetics, not leaving translational research for SLE as an exception. In fact, as shown recently, role of type I interferon in NPSLE has been suggested, where this inflammatory cytokine essential in SLE may activate microglia and lead to synapse loss in lupus prone mice [50]. Similar type I interferon activation was observed in patients' brain. Our cognitions and moods, in other words, thoughts and feelings might be recognized as complex products of neurotransmitters, hormones, cytokines, viability of neuronal cells and low-voltage electricity in our brain. Therefore, we believe there will be a better way to understand the complexity of the pathogenesis in NPSLE and to treat suffering patients for their better prognosis, cognitive/psychological functions and qualities of life.

Acknowledgements I appreciate supports and advice from Drs. Yuichiro Fujieda, Kenji Oku, Yuka Shimizu, Naoki Ohnishi, and Tatsuya Atsumi.

References

1. ACR Ad Hoc Committee on Neuropsychiatric Lupus Nomenclature. The American College of Rheumatology nomenclature and case definitions for neuropsychiatric lupus syndromes. Arthritis Rheum. 1999;42:599–608.
2. Unterman A, et al. Neuropsychiatric syndromes in systemic lupus erythematosus: a meta-analysis. Semin Arthritis Rheum. 2011;41:1–11.
3. Sibley JT, et al. The incidence and prognosis of central nervous system disease in systemic lupus erythematosus. J Rheumatol. 1992;19:47–52.
4. Karassa FB, et al. Predictors of clinical outcome and radiologic progression in patients with neuropsychiatric manifestations of systemic lupus erythematosus. Am J Med. 2000;109:628–34.
5. Ginzler EM, et al. A multicenter study of outcome in systemic lupus erythematosus. I. Entry variables as predictors of prognosis. Arthritis Rheum. 1982;25:601–11.
6. Jonsen A, et al. Outcome of neuropsychiatric systemic lupus erythematosus within a defined Swedish population: increased morbidity but low mortality. Rheumatology. 2002;41:1308–12.
7. Hanly JG, et al. Prospective analysis of neuropsychiatric events in an international disease inception cohort of patients with systemic lupus erythematosus. Ann Rheum Dis. 2010;69:529–35.
8. Zirkzee EJ, et al. Mortality in neuropsychiatric systemic lupus erythematosus (NPSLE). Lupus. 2014;23:31–8.
9. Arinuma Y, et al. Brain MRI in patients with diffuse psychiatric/neuropsychological syndromes in systemic lupus erythematosus. Lupus Sci Med. 2014;1:e000050.
10. Shimizu Y, et al. Post-steroid neuropsychiatric manifestations are significantly more frequent in SLE compared with other systemic autoimmune diseases and predict better prognosis compared with de novo neuropsychiatric SLE. Autoimmun Rev. 2016;15:786–94.
11. Kono M, et al. Long-term outcome in Japanese patients with lupus nephritis. Lupus. 2014;23:1124–32.
12. Mok CC, et al. Effect of renal disease on the standardized mortality ratio and life expectancy of patients with systemic lupus erythematosus. Arthritis Rheum. 2013;65:2154–60.
13. Yap DY, Tang CS, Ma MK, Lam MF, Chan TM. Survival analysis and causes of mortality in patients with lupus nephritis. Nephrol Dial Transplant. 2012;27:3248–54.
14. Shinjo SK, et al. Antimalarial treatment may have a time-dependent effect on lupus survival: data from a multinational Latin American inception cohort. Arthritis Rheum. 2010;62:855–62.
15. Monahan RC, et al. Neuropsychiatric symptoms in systemic lupus erythematosus: impact on quality of life. Lupus. 2017;26:1252–9.
16. Magro-Checa C, et al. Outcomes of neuropsychiatric events in systemic lupus erythematosus based on clinical phenotypes; prospective data from the Leiden NP SLE cohort. Lupus. 2017;26:543–51.
17. Gladman D, et al. The development and initial validation of the Systemic Lupus International Collaborating Clinics/American College of Rheumatology damage index for systemic lupus erythematosus. Arthritis Rheum. 1996;39:363–9.
18. Magro-Checa C, et al. Changes in white matter microstructure suggest an inflammatory origin of neuropsychiatric systemic lupus erythematosus. Arthritis Rheum. 2016;68:1945–54.
19. Zirkzee EJ, et al. Prospective study of clinical phenotypes in neuropsychiatric systemic lupus erythematosus; multidisciplinary approach to diagnosis and therapy. J Rheumatol. 2012;39:2118–26.
20. Fujieda Y, et al. Predominant prevalence of arterial thrombosis in Japanese patients with antiphospholipid syndrome. Lupus. 2012;21(14):1506.
21. Cervera R, et al. Antiphospholipid syndrome: clinical and immunologic manifestations and patterns of disease expression in a cohort of 1,000 patients. Arthritis Rheum. 2002;46:1019–27.
22. Bernatsky S, et al. Mortality related to cerebrovascular disease in systemic lupus erythematosus. Lupus. 2006;15:835–9.

23. Otomo K, et al. Efficacy of the antiphospholipid score for the diagnosis of antiphospholipid syndrome and its predictive value for thrombotic events. Arthritis Rheum. 2012;64:504–12.
24. Sciascia S, et al. GAPSS: the Global Anti-Phospholipid Syndrome Score. Rheumatology. 2013;52:1397–403.
25. Oku K, et al. How to identify high-risk APS patients: clinical utility and predictive values of validated scores. Curr Rheumatol Rep. 2017;19:51.
26. Hanly JG. Diagnosis and management of neuropsychiatric SLE. Nat Rev Rheumatol. 2014;10:338–47.
27. Hanly JG, et al. Screening for cognitive impairment in systemic lupus erythematosus. J Rheumatol. 2012;39(7):1371.
28. Denburg SD, Denburg JA. Cognitive dysfunction and antiphospholipid antibodies in systemic lupus erythematosus. Lupus. 2003;12:883–90.
29. Hanly JG, et al. Cognitive function in systemic lupus erythematosus: results of a 5-year prospective study. Arthritis Rheum. 1997;40:1542–3.
30. Waterloo K, et al. Neuropsychological function in systemic lupus erythematosus: a five-year longitudinal study. Rheumatology. 2002;41:411–5.
31. Ainiala H, et al. Validity of the new American College of Rheumatology criteria for neuropsychiatric lupus syndromes: a population-based evaluation. Arthritis Rheum. 2001;45:419–23.
32. Hanly JG, et al. Neuropsychiatric events in systemic lupus erythematosus: attribution and clinical significance. J Rheumatol. 2004;31:2156–62.
33. Sanna G, et al. Neuropsychiatric manifestations in systemic lupus erythematosus: prevalence and association with antiphospholipid antibodies. J Rheumatol. 2003;30:985–92.
34. Hirohata S, et al. Accuracy of cerebrospinal fluid IL-6 testing for diagnosis of lupus psychosis. A multicenter retrospective study. Clin Rheumatol. 2009;28:1319–23.
35. Abe G, et al. Brain MRI in patients with acute confusional state of diffuse psychiatric/neuropsychological syndromes in systemic lupus erythematosus. Mod Rheumatol. 2017;27:278–83.
36. Barile-Fabris L, et al. Controlled clinical trial of IV cyclophosphamide versus IV methylprednisolone in severe neurological manifestations in systemic lupus erythematosus. Ann Rheum Dis. 2005;64:620–5.
37. Neuwelt CM, et al. Role of intravenous cyclophosphamide in the treatment of severe neuropsychiatric systemic lupus erythematosus. Am J Med. 1995;98:32–41.
38. Stojanovich L, et al. Neuropsychiatric lupus favourable response to low dose i.v. cyclophosphamide and prednisolone (pilot study). Lupus. 2003;12:3–7.
39. Neuwelt CM. The role of plasmapheresis in the treatment of severe central nervous system neuropsychiatric systemic lupus erythematosus. Ther Apher Dial. 2003;7:173–82.
40. Camara I, et al. Treatment with intravenous immunoglobulins in systemic lupus erythematosus: a series of 52 patients from a single centre. Clin Exp Rheumatol. 2014;32:41–7.
41. Cobo-Ibanez T, et al. Efficacy and safety of rituximab in the treatment of non-renal systemic lupus erythematosus: a systematic review. Semin Arthritis Rheum. 2014;44:175–85.
42. Tokunaga M, et al. Efficacy of rituximab (anti-CD20) for refractory systemic lupus erythematosus involving the central nervous system. Ann Rheum Dis. 2007;66:470–5.
43. Ginzler E, et al. Long-term maintenance therapy with azathioprine in systemic lupus erythematosus. Arthritis Rheum. 1975;18:27–34.
44. Bertsias GK, et al. EULAR recommendations for the management of systemic lupus erythematosus with neuropsychiatric manifestations: report of a task force of the EULAR standing committee for clinical affairs. Ann Rheum Dis. 2010;69:2074–82.
45. Khamashta MA, et al. The management of thrombosis in the antiphospholipid-antibody syndrome. N Engl J Med. 1995;332:993–7.
46. Crowther MA, et al. A comparison of two intensities of warfarin for the prevention of recurrent thrombosis in patients with the antiphospholipid antibody syndrome. N Engl J Med. 2003;349:1133–8.

47. Finazzi G, et al. A randomized clinical trial of high-intensity warfarin vs. conventional antithrombotic therapy for the prevention of recurrent thrombosis in patients with the antiphospholipid syndrome (WAPS). J Thromb Haemost. 2005;3:848–53.
48. Cohen H, et al. Rivaroxaban versus warfarin to treat patients with thrombotic antiphospholipid syndrome, with or without systemic lupus erythematosus (RAPS): a randomised, controlled, open-label, phase 2/3, non-inferiority trial. Lancet Haematol. 2016;3:e426–36.
49. Chau SY, Mok CC. Factors predictive of corticosteroid psychosis in patients with systemic lupus erythematosus. Neurology. 2003;61:104–7.
50. Bethea HN, et al. Redirecting the substrate specificity of heparan sulfate 2-O-sulfotransferase by structurally guided mutagenesis. Proc Natl Acad Sci U S A. 2008;105:18724–9.

Index

A
ACR 1999 nomenclature, 60
Acute confusional state (ACS), 32, 46, 62, 83, 105, 144, 160, 171, 177
Acute confusional state (delirium), 134
Acute organic brain syndrome, 135
Acute pulmonary hemorrhage, 47
Adhesion molecules, 35
Aicardi-Goutières syndrome (AGS), 20
Anti-aquaporin 4 (AQP4) antibodies, 7, 34, 51, 118
Anti-asialo GM1, 34
Anti-β2GP1, 30
Anti-BLyS antibodies (Belimumab, Tabalumab and Blisibimod), 164
Antibody-dependent cellular toxicity, 162
Anti-cardiolipin, 30, 45, 145
Anti-CD22 antibodies (Epratuzumab), 163
Anticoagulation, 178
Anti-dsDNA antibodies, 97
Anti-endothelial cell antibodies, 166
Antiepileptic drugs, 146
Antigenic target, 107
Anti-glycolipid antibodies, 34
Anti-IFN-α antibodies (Sifalimumab and Rontalizumab), 165
Anti-IL-6 receptor antibodies (Tocilizumab), 165
Anti-neuronal antibodies, 33, 99
Anti-NMDR NR2 (anti-NR2) antibodies, 32, 49
Anti-NR2 glutamate receptor, 83, 97
Antinuclear antibody (ANA), 96
Antiphospholipid antibodies (aPL), 6, 45, 95, 120, 136, 143, 172
Anti-phospholipid syndrome (APS), 175
Anti-platelet agents, 9, 143, 147, 178
Anti-ribosomal P protein antibodies (anti-P), 31, 83, 97, 133
Anti-RNP, 33
Anti-Sm antibodies, 32, 97
Anti-terminal complment component C5a and C5b-9 antibodies (Eculizumab), 165
Antithrombotic drugs, 147
Anxiety disorder (AD), 63, 134, 145
aPL-score, 175
APOE, 23
A proliferation-inducing ligand (APRIL), 83, 164
Apoptosis, 31
Apoptotic cells, 21
Apparent diffusion coefficient (ADC) mapping, 117
Aquaporin 4, 107
Arteritis, 55
Aseptic meningitis, 66
Aspirin, 147
Astrocytes, 31, 34
Atrophic lesion, 122
Atrophy, 44
Attention, 135
Autoantibodies, 29, 44, 55
Autoantibody production, 163
Autonomic disorder, 71
Azathioprine, 150, 178

B
B-cell activating factor of TNF family (BAFF), 83, 164
B-cell-specific antigen CD20, 156
B-cell-triggered apoptosis, 162

Belimumab, 150, 151
BLK, 22
Blood-brain barrier (BBB), 30, 44, 46, 78, 137, 163
 damages, 48
 dysfunction, 35
Blood chemistry tests, 97
Blood flow, 124
B lymphocyte stimulator (BLyS), 164
BLyS and APRIL receptor (Atacicept), 164
Brain atrophy, 44
Brainstem, 45, 47

C
C1q, 36
C3, 36, 53
C4, 20
C4d, 45
C5a, 36
C5a receptor (C5aR), 36
C5b-9, 45
Calcification, 54
Catatonia, 136
Cause of death, 172
CD19+ B cells, 161
CD86, 37, 54
CD244, 23
Central and peripheral nervous system involvement, 94
Central nervous system (CNS), 10, 114
Cerebritis, 34
Cerebrovascular disease (CVD), 10, 44, 45, 50, 67, 103, 147, 170, 174
Chemokines, 88, 164
Chemokines as biomarkers, 85–88
Chorea, 104
Choroid plexus, 53
Claudin 5, 36
Clearance, 21
Clopidogrel, 147
CNS lymphoma, 119
Coagulation necrosis, 54, 55
Cognitive dysfunction (CD), 7, 9, 30, 44, 49, 64, 104, 135, 136, 145, 162, 172, 175
Complement, 21, 36, 45
Complement (C3/C4, or CH50) determinations, 97
Complement-mediated cellular lysis, 162
Computed tomography (CT), 114

Cortical atrophy, 49
Corticosteroid (CS)-induced psychosis, 179
Corticosteroid-induced psychiatric disorder (CIPD), 137, 138
Corticosteroids, 147, 148
Cranial neuropathy, 50, 71, 106
CSF anti-neuronal antibody titers, 160
CSF anti-NR2, 32
CSF IgG index, 34, 83
CSF IgG levels, 99
CSF interleukin-6 levels, 8, 160
CSF Q-albumin ratio, 99
Cyclophosphamide, 148, 156
Cytokine/chemokine storm, 90
Cytokines, 49, 164
Cytokines as biomarkers, 79, 81–84

D
Damaged BBB, 163
Delirium, 134
Demyelinating syndrome, 8, 67
Demyelination, 52
De novo NPSLE, 179
Depletion of B cells, 163
Depression, 31, 133
Diagnostic and statistical manual of mental disorders, 4th edition (DSM-IV), 132
Diffuse and focal neuropsychiatric events, 94
Diffuse NP events, 174
Diffuse NPSLE, 32–33, 44, 49
Diffuse psychiatric/neuropsychological syndrome (diffuse NPSLE), 3, 60, 116, 142
Diffusion tensor imaging (DTI), 123
Diffusion-weighted imaging (DWI), 100, 116
Direct oral anticoagulants (DOACs), 179
DNASE1, 21
DNASE1L3, 21
Dopamine antagonists, 147
Dual antiplatelet therapy (DAPT), 179
Dying cell, 21

E
Electroconvulsive therapy, 136
Electroencephalography (EEG), 100
Endothelial cells, 30, 35, 45
Ethnicity, 6
Exploratory phase II/III SLE evaluation of rituximab (EXPLORER) trial, 157

Index

F
FCGR2A, 23
FCGR3A, 23
FCGR3B, 23
Fluid attenuated inversion recovery imaging (FLAIR), 116
Focal NP events, 174
Focal NPSLE, 33, 43
Fractalkine/CX3CL1 (a ligand of CX3CR1d), 87

G
Gadolinium (Gd)-enhancement, 117
Genetic epidemiology, 15
Genetics, 15–24
Genome-wide association studies (GWAS), 16
Gliosis, 44
Global anti-phospholipid syndrome score (GAPSS), 175
Glucocorticoids, 157
Good evidence, 108
Granulocyte-colony stimulating factors (G-CSF) levels, 84
Gray matter, 52
Gray matter hyperintensity (GMH), 121
Grey matter high intensity lesions (GMHIs), 44
GTF2I, 22
GTF2IRD1, 22
Guillain-Barré syndrome, 53, 71

H
Headache, 9, 68, 101, 146
Hemoglobin, 35
High rate of efficacy, 163
Hippocampus, 31, 47
HLA-DRB1, 16
Hydroxychloroquine, 143, 149
Hypertrophic pachymeningitis, 122
Hypoperfusion, 44

I
IFIH1, 20, 21
IKBKE, 22
IL-6 mRNA, 31, 47
IL-8/CXCL8 (a ligand of CXCR1 and CXCR2), 86
Immune complexes, 45, 81
Immunosuppressants, 143
Immunosuppressive therapy, 10

Inflammatory NPSLE, 49, 173
Interferon-α (IFN-α), 37, 81
Interferon-gamma inducible protein-10/ CXCL10 (a ligand of CXCR3), 87
Interleukin-6 (IL-6), 82–84, 100, 133
Interleukin-8 (IL-8), 10, 30, 32
Interleukin-10 (IL-10), 81
Intrathecal Ig production, 34
Intrathecal ratio of IP-10 to MCP-1, 100
Intravenous cyclophosphamide (IVCY), 162, 177
Intravenous (IV) immunoglobulin, 149
Intravenous pulses of methylprednisolone, 162
IP-10, 10
IP-10/MCP-1 ratio, 88
IRF5, 20
IRF8, 20
Irreversible damages, 52, 172
Ischemic NPSLE, 173

J
John Cunningham virus (JCV), 108

L
Liquefaction necrosis, 52
Low-dose aspirin, 143
Low-dose IVCY, 178
Lupus anticoagulant (LA), 30, 45
Lupus headache, 7, 146
Lupus nephritis assessment with rituximab (LUNAR) trial, 156–157
Lupus psychosis, 4, 31, 105, 148

M
Magnetic resonance imaging (MRI), 44, 46, 48, 100, 116
Magnetic resonance spectroscopy (MRS), 123
Magnetization transfer imaging (MTI), 49, 123, 173
Major depression, 133
Major histocompatibility complex (MHC), 16, 19, 20
Mania, 134
Mechanism of action of rituximab, 162
Memory, 135
Meta-analyses, 4, 170
Methylprednisolone, 147
MHC class II antigen, 37, 54
Microangiopathy, 46
Microglia, 36, 53

Microhaemorrhages, 44, 49, 55
Micro-infarction, 55
Microscopic infarcts, 45
Microthrombi, 44, 45, 54
Microtubule-associated protein 2 (MAP-2), 34
Microvasculopathy, 45
Microvessel occlusion, 55
Mixed connective tissue disease, 96
Molecular mechanisms, 89
Monocyte chemoattractant protein-1/ CCL2 (a ligand of CCR2), 85, 86
Monocytes, 31, 35
Mononeuritis multiplex, 53
Mononeuropathy (single/multiplex), 71
Mood disorder (MD), 7, 9, 32, 64, 133, 145, 170
Mortality rate, 9, 170
Movement disorder, 146
Movement disorder (chorea), 69
MRI with Gd-enhacement, 122
MRL/lpr mice, 36
Multidisciplinary evaluation, 173
Mutism, 160
MX1, 37
Myasthenia gravis (MG), 72
Mycophenolate mofetil (MMF), 119, 149, 150, 156
Myelopathy, 8, 51, 52, 69

N
NCF1, 22
Neuorogenesis, 84
Neuroblastoma, 31
Neurologic syndromes, 60, 66, 116, 142
Neuromyelitis optica (NMO), 107
Neuromyelitis optica spectrum disorders (NMOSD), 7, 34, 51, 67, 118
Neuronal connectivity, 36
Neuronal surface P antigens (NSPAs), 31
Neurons, 30, 47
Neuropsychiatric systemic lupus erythematosus (NPSLE), 43, 60, 113
  EULAR recommendations, management, 101
  guidelines for diagnosis, 101
  nomenclature and case definition, 170
Neuropsychologic tests, 10
Neurotrophic factor, 84
Neutrophil extracellular traps (NETs), 21
Neutrophils, 21
NF-κB, 22, 32, 36, 83
NMDAR NR1 subunit, 33

N-methyl-D-aspartate (NMDA) receptors, 32
NMO-IgG, 52
NMO IgG/anti-aquaporin 4 antibodies, 97
NMO-specific autoantibodies, 107
Nucleic acid, 20, 21

O
Occludin, 36
Oligoclonal bands, 99
Oligoclonal IgG bands, 35
Organic brain syndrome, 2, 32, 144
Ovarian teratoma, 32

P
Paraplegia, 52
Pathogenesis, 89
Pathogenetic role for autoantibodies, 88
Pathogenic factors, 88, 89
Pathophysiology, 89
Pediatric SLE, 8
Peripheral nerve, 53
Peripheral nervous system (PNS), 10
Perivascular lymphocytic infiltrates, 44
Perivenous changes, 54, 55
Plasmacytoid dendritic cells, 81
Plasmapheresis, 149
Plexopathy, 72
PNS disorders, 106
Polyarteritis nodosa, 53
Polyneuropathy, 72
Positron emission tomography (PET), 124
Posterior reversible encephalopathy syndrome (PRES), 118
Prednisolone, 147
Prevalence, 4, 169
Primary NPSLE, 130
Prognosis, 62
Progressive multifocal leukoencephalopathy (PML), 108
Proinflammatory cytokines and chemokines, 78
Prospective study, 6
Proteomics, 34
Psychiatric disorders, 132
Psychiatric/neuropsychological manifestations, 3
Psychiatric symptoms, 130
Psychomotor speed, 135
Psychosis, 2, 65, 66, 132, 145

Index

Q
Q albumin, 33

R
Reactive microglia, astrocytes and pericytes, 78
Regulated upon activation, normal T-cell expressed and secreted (RANTES)/CCL5, 86
Retrospective studies, 10
Reversible cerebral vasoconstriction syndrome (RCVS), 119
Reversible focal neurological deficits, 7, 50
Reversible posterior leukoencephalopathy syndrome (RPLS), 107
Risk factors, 6, 94, 95
Rituximab, 150, 160
RNASEH2A/2B/2C, 21
Routine CSF tests, 99

S
SAMHD1, 21
Secondary NPSLE, 130
Secondary prophylaxis, 178
Seizure disorders, 70, 104
Seizures, 2, 44, 50, 70, 146, 170
Sequencing, 20
Serum anti-Sm, 33
Signal transduction pathways, 89
Since anti-NR2, 52
Single nucleotide polymorphisms (SNPs), 16
Single photon emission cerebral tomography (SPECT), 124
Sjogren's syndrome, 96
SLE disease activity, 95
SLE disease activity index (SLEDAI), 95
SLICC/ACR damage index, 173
SLICC inception cohort, 6
Small-fiber neuropathies, 53
Soluble ICAM-1, 35
Specialized endothelial cells (ECs), 78
Standardized mortality ratio (SMR), 9, 171
STAT4, 22
Steroid induced psychosis, 8, 148
Steroid psychosis, 8
Suicide, 136, 137
Symptomatic treatment, 144
Synapse loss, 53
Synaptic function, 32
Synaptic refinement, 36
Syndrome, 2

T
Thromboembolic events, 143
Thrombotic, 45
1,000 mg doses, 160
TLR4, 30
TLR7, 20
TNFAIP3, 22
TNF-α, 31, 79
TNIP1, 22
Transcription factors, 89
Transverse myelitis, 7, 47, 51, 105
TREX1, 21, 22
T1-weighted image, 116
T2-weighted image, 116
TYK2, 20
Type I interferon (IFN), 20, 21, 36, 53, 180
Type I interferonopathies, 20

U
UBE2L3, 22

V
Vasculitis, 7, 44, 47, 50, 117
Vasculopathy, 44, 49
Visual and auditory hallucinations, 162

W
Warfarin, 147, 179
White matter, 52, 136
White matter high intensity lesions (WMHIs), 44
White matter hyperintensity (WMH), 119, 121
White matter lesions, 49
Working incapability, 174

Z
ZO-1, 36

Printed by Printforce, the Netherlands